THE DICTIONARY OF BODY LANGUAGE

Also by Joe Navarro

What Every BODY Is Saying: An Ex-FBI Agent's Guide to Speed-Reading People

Louder Than Words: Take Your Career from Average to Exceptional with the Hidden Power of Nonverbal Intelligence

Phil Hellmuth Presents Read 'Em and Reap: A Career FBI Agent's Guide to Decoding Poker Tells

THE DICTIONARY OF BODY LANGUAGE

A Field Guide to Human Behavior

JOE NAVARRO

wm

WILLIAM MORROW

An Imprint of HarperCollins*Publishers*

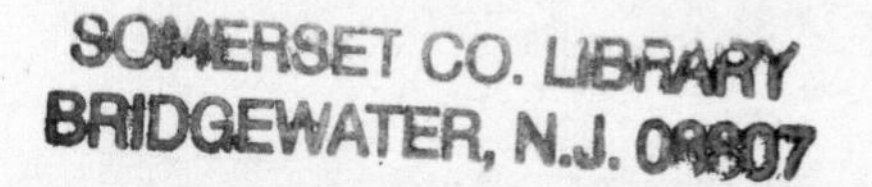

While the manuscript for this book was reviewed by the Federal Bureau of Investigation (FBI) prior to publication, the opinions and thoughts expressed herein are those of the author exclusively.

 For information, address HarperCollins Publishers, 195 Broadway, New York, NY 10007.

HarperCollins books may be purchased for educational, business, or sales promotional use. For information, please e-mail the Special Markets Department at SPsales@harpercollins.com.

FIRST EDITION

Designed by Diahann Sturge

Illustration credits: pages 9, 27, 53, 85, 91, 97, 105, 119, 145, 171 © Anna Rassadnikova/Shutterstock, Inc.; page 109 © freelanceartist/Shutterstock, Inc.; page 155 © Oksana Usenko/Shutterstock, Inc.; page 161 © makar/Shutterstock, Inc.

Library of Congress Cataloging-in-Publication Data has been applied for.

ISBN 978-0-06-284687-7

18 19 20 21 22 LSC 10 9 8 7 6 5 4 3 2 1

This book is dedicated to the love of my life,
my best friend, and the first editor of everything
I do—my wife, Thryth Hillary Navarro

If language was given to men to conceal their thoughts, then gesture's purpose was to disclose them.

—John Napier

CONTENTS

INTRODUCTION

In 1971, at the age of seventeen, for reasons unknown to me then or now, I began to keep a journal on human behavior. I catalogued all sorts of "nonverbals"—what is more generally called body language. At first it was the quirky things people did: why did they roll their eyes when they were disbelieving or reach for their neck when they heard bad news? Later it became more nuanced: why did women play with their hair while on the phone or arch their eyebrows when they greeted one another? These were small actions, but they captured my curiosity. Why did humans do such things, in such variety? What was the purpose of these behaviors?

I admit it was an odd pursuit for a teenager. My friends told me as much; they were focused on trading baseball cards, knowing who had the best batting average or kicked the most extra points that season. I was far more interested in learning the intricacies of human behavior.

In the beginning I catalogued my observations on three-by-five-inch cards for my own benefit. At that time I was unfamiliar with the work of Charles Darwin, Bronisław Malinowski, Edward T. Hall, Desmond Morris, or my future friend Dr. David Givens—the giants in the field of human behavior. I was simply interested in how others acted, and *why,* and I wanted to preserve my observations. I never thought I would still be collecting them on index cards forty years later.

Over the years, I collected several thousand entries. Little did I know back then that I would later become an FBI Special Agent and would, for the next twenty-five years, use those observations as I pursued criminals, spies, and terrorists. But perhaps, given my interest in how and why people behave, that was the natural trajectory all along.

I CAME TO the United States as a refugee fleeing Communist-controlled Cuba. I was eight years old and didn't speak English. I had to adjust quickly—in other words, I had to observe and decode my new surroundings. What native speakers took for granted, I could not. My new existence consisted of deciphering the only thing that made sense—body language. Through their countenance, their look, the softness in their eyes, or the tension in their face, I learned to interpret what others implied. I could figure out who liked me, who was indifferent toward my existence, whether someone was angry or upset with me. In a strange land, I survived by observing. There was no other way.

Of course, American body language was a little different

from Cuban body language. People in America spoke with a different cadence and vibrancy. Cubans got close to one another when they spoke, and often touched. In America they stood farther apart, and social touching might receive an uncomfortable glance or worse.

My parents worked three jobs each, so they did not have the time to teach me these things—–I had to learn them on my own. I was learning about culture and the influence it has on nonverbals, even if I couldn't have put it in those words at the time. But I did know that some behaviors were different here, and I had to understand them. I developed my own form of scientific inquiry, observing dispassionately and validating everything I saw not once or twice but many times before it made its way onto an index card. As my cards grew in number, certain patterns in behavior began to stand out. For one, most behaviors could be broadly categorized as markers of either psychological comfort or discomfort; our bodies reveal very accurately, in real time, our state of unease.

I would later learn that many of these *comfort markers* or behaviors, to be more precise, originated in the mammalian or emotional areas of the brain—what is often referred to as the limbic system. This type of involuntary response squared with what I had seen in Cuba and was seeing now in America. At school or through the window at the corner store, people would flash their eyes with their eyebrows to greet those they truly liked. Such universal behaviors I grew to trust as authentic and reliable. What I did doubt was the spoken word. How often, after I had learned English, I heard people say they liked something when just an instant earlier I had seen their face reveal the complete opposite.

And so, too, I learned at an early age about deception. People often lie, but their nonverbals usually reveal how they actually feel. Children, of course, are terrible liars; they might nod to acknowledge they have done something bad even as they are verbally denying it. As we get older, we get better at lying, but a trained observer can still spot the signs that say something is wrong, there are issues here, a person does not appear to be completely forthcoming, or someone lacks confidence in what he is saying. Many of those signals or behaviors are collected here in this book.

As I grew older, I came to rely more and more on nonverbals. I relied on them at school, in sports, in everything I did—even playing with my friends. By the time I had graduated from Brigham Young University, I had collected more than a decade's worth of observations. There, for the first time, I was living among many more cultures (east Europeans, Africans, Pacific Islanders, Native Americans, Chinese, Vietnamese, and Japanese, among others) than I had seen in Miami, and this allowed me to make further observations.

At school I also began to discover the fascinating scientific underpinnings of many of these behaviors. To take just one example: in 1974 I got to see congenitally blind children playing together. It took my breath away. These children had never seen other children yet were exhibiting behaviors that I had thought were visually learned. They were demonstrating "happy feet" and the "steeple" with their hands, despite having never witnessed them. This meant these behaviors were hardwired into our DNA, part of our paleo-circuits—these very ancient circuits that ensure our survival and ability to communicate and are thus universal. Throughout my college career, I learned about the evolutionary basis of many of these behaviors, and throughout this book, I will reveal these often surprising facts we take for granted.

When I finished my studies at Brigham Young University, I received a phone call asking me to apply to the FBI. I thought it was a joke, but the next day two men in suits knocked on my door and handed me an application and my life changed forever. In those days, it was not unusual for FBI scouts to look for talent on campus. Why my name was handed up, or by who, I never learned. I can tell you that I was more than elated to be asked to join the most prestigious law enforcement agency in the world.

I was the second-youngest agent ever hired by the FBI. At the age of twenty-three I had again entered a new world. Though I felt unprepared in many ways to be an agent, there was one domain I had mastered: nonverbal communication. This was the only area where I felt confident. FBI work is, for the most part, about making observations. Yes, there are crime scenes to process and criminals to apprehend, but the majority of the job is talking to people, surveilling criminals, conducting interviews. And for that I was ready.

My career in the FBI spanned twenty-five years, the last thirteen of which I spent in the Bureau's elite National Security Behavioral Analysis Program (NS-BAP). It was in this unit, designed to analyze the top national security cases, that I got to utilize my nonverbal skills as if on steroids. This unit, comprising just six agents selected from among twelve thousand FBI Special Agents, had to achieve the impossible: identify spies, moles, and hostile intelligence officers seeking to do harm to the United States under diplomatic cover.

During my time in the field I honed my understanding of body language. What I observed could never be replicated in a university laboratory. When I read scientific journals about deception and body language, I could tell that the authors had never actually interviewed a psychopath, a terrorist, a "made" Mafia member, or an intelligence officer from the Soviet KGB. Their findings might be true in a lab setting, using university students. But they understood little of the real world. No lab could replicate what I had observed in vivo, and no researcher could approximate the more than thirteen thousand interviews I had done in my career, the thousands of hours of surveillance video I had observed, and the behavioral notations that I had made. Twenty-five years in the FBI was my graduate school; putting multiple spies in prison based on nonverbal communications was my dissertation.

After retiring from the FBI, I wanted to share what I knew about body language with others. *What Every BODY Is Saying,* published in 2008, was the product of that quest. In that book the concepts of "comfort" and "discomfort" took center stage, and I unveiled the ubiquity of "pacifiers"—such as touching our faces or stroking our hair—body behaviors we use to deal with everyday stress. I also sought to explain where these universal behaviors came from, drawing upon psychological research, evolutionary biology, and cultural contexts to explain *why* we do the things we do.

What Every BODY Is Saying became an international best seller; it has been translated into dozens of languages and has sold more than a million copies around the world. When I wrote *What Every BODY Is Saying,* I had no idea how popular it would become. At my speaking engagements in the years following its publication, I kept hearing the same thing: people wanted more, and they wanted it in a more easily accessible format. What many readers asked for was a field guide of sorts, a quick reference manual for behaviors they might encounter in day-to-day life.

The Dictionary of Body Language is that field guide. Organized by areas of the body—moving from the head down to the feet—it contains more than four hundred of the most important body-language observations I have made over the course of my career. My hope is that reading through *The Dictionary of Body Language* will give you the same insight into human behavior that I and other FBI agents have used to decode human behavior. Of course, we have used it when questioning suspects of crime. But you can use it as I have every day since I came to this country—to more fully understand those we interact with at work or at play. In social relationships, I can think of no better way to comprehend your friends or partners than by studying the primary means by which we communicate—nonverbally.

If you have ever wondered why we do the things we do, or what a particular behavior means, my hope is to satisfy your curiosity. As you go through the dictionary, act out the behaviors that you read about and get a sense for how they appear as well as they feel. By acting these out, you will better remember them the next time you see them. If you are like me and enjoy people watching, if you want to discern what people are thinking, feeling, desiring, fearing, or intending, whether at work, at home, or in the classroom, read on.

THE HEAD

All behavior, of course, originates from inside the head. The brain is constantly at work, whether on a conscious or subconscious level. The signals that go out from the brain regulate the heart, breathing, digestion, and many other functions—but the exterior of the head is tremendously important as well. The hair, forehead, eyebrows, eyes, nose, lips, ears, and chin all communicate in their own way—from our general health to emotional distress. And so we begin with the part of the body that, from the time we are born until we die, we look to for useful information—first as parents, later as friends, work mates, lovers—to reveal for us what is in the mind.

1. **HEAD ADORNMENT**—Head adornment is used across all cultures for a variety of reasons. It can communicate leadership status (Native American chiefs' feather headdresses), occupation (a hard hat or miner's hat), social status (a bowler hat or an Yves Saint Laurent pillbox hat), hobbies (bicycle or rock-climbing helmet),

religion (cardinal's cap, Jewish yarmulke), or allegiance (favorite sports team, labor union). Head adornments may offer insight into individuals: where they fit in society, their allegiances, their socioeconomic status, what they believe, how they see themselves, or even the degree to which they defy convention.

2. **HAIR**—Sitting conveniently on top of the head, our hair conveys so much when it comes to nonverbal communication. Healthy hair is something all humans look for, even on a subconscious level. Hair that is dirty, unkempt, pulled out, or uncared for may suggest poor health or even mental illness. Hair attracts, entices, conforms, repels, or shocks. It can even communicate something about our careers; as renowned anthropologist David Givens puts it, hair often serves as an "unofficial résumé," revealing where one ranks in an organization. And in many cultures hair is critical to dating and romance. People tend to follow both cultural norms and current trends with their hair; if they ignore these societal standards, they stand out.

3. **PLAYING WITH HAIR**—Playing with our hair (twirling, twisting, stroking) is a *pacifying behavior*. It is most frequently utilized by women and might indicate either a good mood (while reading or relaxing) or stress (when waiting for an interview, for example, or experiencing a bumpy flight). Note that when the *palm*

of the hand faces the head it is more likely to be a pacifier, as opposed to the palm-out orientation discussed below. Pacifying behaviors soothe us psychologically when we feel stress or anxiety; they also help us to pass the time. As we grow older we go from pacifying by sucking our thumbs to such behaviors as lip biting, nail biting, or facial stroking.

4. **PLAYING WITH HAIR (PALM OUT)**—When women play with their hair with the *palm of the hand facing out*, it is more of a public display of comfort—a sign that they are content and confident around others. We usually only expose the underside of our wrists to others when we are comfortable or at ease. This is often seen in dating scenarios where the woman will play with her hair, palm out, while talking to someone in whom she is interested.

5. **RUNNING FINGERS THROUGH HAIR (MEN)**—When stressed, men will run their fingers through their hair both to ventilate their heads (this lets air in to cool the vascular surface of the scalp) and to stimulate the nerves of the skin as they press down. This can also be a sign of concern or doubt.

6. **VENTILATING HAIR (WOMEN)**—The ventilating of hair is a powerful pacifier, relieving both heat and stress. Women ventilate their hair differently than

men. Women lift up the hair at the back of their neck quickly when concerned, upset, stressed, or flustered. If they do it repeatedly, most likely they are overly stressed. Nevertheless, we cannot discount overheating due to physical activity or ambient temperature as a cause. Men tend to ventilate on the top of the head by running their fingers through the hair.

7. **HAIR FLIPPING/TOUCHING**—Hair flipping, touching, or pulling is common when we are trying to attract the attention of a potential mate. The movement of the hand as it touches the hair is often deemed attractive (note most any hair commercial). Our *orientation reflex* (OR), a primitive reaction that alerts us to any movement, is especially attuned to hand movements—something magicians have always counted on. A hand reaching for the hair can draw our attention even from across the room. Incidentally, the orientation reflex operates on such a subconscious level, it is even seen in coma patients as the eyes track movement.

8. **HAIR PULLING**—The intentional and repetitive pulling out of hair is called *trichotillomania*. Hair pulling is more often seen in children and teenagers who are experiencing stress, but it is also occasionally seen in adults. Men tend to pluck hair from the corners of their eyebrows, while women are far more wide-ranging: plucking their eyelids, head hair, eyebrows,

and arm hair. This is a stress response; even birds will pull out their own feathers when stressed. The repetitive pulling out of the hair, like a nervous tic, pacifies by stimulating nerve endings; unfortunately, when it becomes severe, it requires medical intervention.

9. **HEAD NODDING**—During conversations nodding serves to affirm, usually in cadence, that the person is hearing and receptive to a message. Generally, it signals agreement, except in those situations where the head nodding is accompanied by lip pursing (see #154), which might suggest disagreement.

10. **HEAD NODDING (CONTRADICTION)**—We usually see this in young children, as when a parent asks a child "Did you break the lamp?" and the child answers "No" but nods. This contradictory behavior betrays the truth. I have seen this with kids, teenagers, and even adults.

11. **HEAD PATTING, BACK OF HEAD**—When we are perplexed or mentally conflicted, we often find ourselves patting the back of our head with one hand, perhaps even stroking our hair downward as we struggle for an answer. This behavior is soothing because of both the tactile sensation and the warmth that is generated. Like most hand-to-body touching, this is a pacifying behavior that reduces stress or anxiety.

12. **HEAD SCRATCHING**—Head scratching soothes us when we have doubts or feel frustrated, stressed, or concerned. You see it with people trying to remember information or when they are perplexed. This explains why it is often seen by teachers as students ponder a test question. Very rapid head scratching often signals high stress or concern. It can also signal the person is conflicted as to what to do next.

13. **HEAD STROKING**—Beyond the function of keeping one's hair in place, people will stroke their hair with the palm of the hand to soothe themselves when stressed or confronted with a dilemma or while pondering how to answer a question. This is not dissimilar to a mother comforting her child by stroking the child's head. This pacifying behavior can have an immediate calming effect. Once more, this behavior may signal doubt or conflict, especially if done to the back of the head.

14. **HEAD SCRATCHING WITH TUMMY RUBBING**—The simultaneous rubbing of the belly and the head indicates doubt or wonder. It can also signal insecurity or incredulity. Interestingly, many primates do this as well.

15. **INTERLACED FINGERS BEHIND HEAD, ELBOWS UP**—The interlacing of the fingers behind the head

with the elbows out is called "hooding" because the person looks like a cobra when it hoods—making the person seem bigger. This is a territorial display we do when comfortable and in charge. When we hood, the interlaced fingers behind the head are both comforting and soothing, while the elbows out project confidence. Hooding is rarely done when someone of higher status is present.

16. **REACHING FOR HEAD (STUPEFIED)**—People who are shocked, in disbelief, or stupefied might suddenly reach for their head with both hands so that the hands are near the ears but not touching them, with the elbows out toward the front. They might hold this position for several seconds as they try to make sense of what happened. This primitive, self-protective response might follow when someone has made a major faux pas, such as a driver crashing into his own mailbox, or a player running toward the wrong goal line.

17. **INTERLACING FINGERS ON TOP OF HEAD**—Usually performed with the palms down, this behavior stands out because it is intended to cover the head and yet the elbows are usually out and wide. We see this when people are overwhelmed, at an impasse, or struggling, when there has been a calamity (after hurricanes or tornados by those who lost property), or when things are not going their way. Note the position of the

elbows: as things get worse, they tend to draw closer together in front of the face almost unnaturally, as if in a vise. Also note the pressure: the worse the situation, the greater the downward pressure of the hands. This behavior is quite different from "hooding" (see #15), where the palms are placed on the back of the head and the person is quite confident.

18. **HAT LIFTING (VENTILATING)**—Under sudden stress, people may suddenly lift up their hat to ventilate their head. This often occurs when receiving bad news, during an argument, or after a heated moment. From a safety perspective, be aware that in situations of high anger (e.g., traffic accidents or road-rage incidents), disrobing (removing hats, shirts, sunglasses) often precedes a fight.

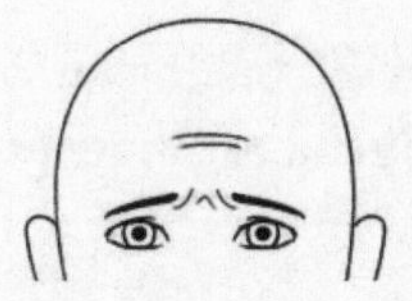

THE FOREHEAD

From the time we are babies, we begin to scan the forehead for information. Even at just a few months of age, infants will respond to the furrows on their mother's forehead—perceiving it as something negative. This small space between the bridge of the nose and the hairline reveals to others, in real time, how we are feeling. It is a remarkable part of the body closely connected to the brain, which allows us to communicate sentiments quickly, accurately, and prominently.

19. FOREHEAD TENSION—On some individuals, stress manifests as sudden tension of the forehead, a result of the stiffening and tensing of underlying muscles. The face has more than twenty distinct muscle groups that can create more than four thousand distinct expressions, according to Dr. Paul Ekman. Six muscles in particular, including the large *occipitofrontalis,* the *procerus,* and the *temporalis,* account for the tightening or furrowing of the forehead when we are stressed. Obviously, one has to see people in a calm environment to get a baseline read on their forehead, but when people are stressed, tension of the forehead is frequently very noticeable and is an excellent indicator that something is wrong.

20. **FOREHEAD FURROWING**—Furrowing of the forehead in response to a stimulus is usually a good indicator that something is amiss, there are issues, or a person is insecure. It is also seen when people are concentrating or trying to make sense of something. Furrowing of the forehead is usually associated with doubt, tension, anxiety, or concern. Keep in mind that Botox, which many use for cosmetic purposes to obscure stress lines on the forehead, might mask true sentiments.

21. **BOTOXED FOREHEAD (ISSUES)**—Both men and women are now taking advantage of Botox injections to erase stress lines on their foreheads. This has created problems for couples and even for children who would normally look to the forehead for information as to how a person might feel. Babies as young as four weeks old will respond to a furrowed forehead as something negative. Interestingly, both children and adults have reported an inability to read their parents or their spouses who have used Botox for emotional cues as easily as they could before.

22. **STRESS LINES**—On some individuals, their life struggles are marked by deep grooves on their forehead, even at a young age. Life experiences often etch our foreheads with lines, furrows, and other indentations. The forehead can reflect a difficult or stressful life or a life that has been spent outdoors in the sun, which tends to make forehead markings more prevalent.

23. **FOREHEAD SWEATING**—If the degree of stress is high enough, some people begin to spontaneously sweat. Sweating is very individual. Some sweat profusely with their first sip of coffee, or climbing a flight of stairs, so make sure to get a baseline of this behavior before jumping to any conclusion. Baseline behaviors are those behaviors we equate with "normal," when a person is not stressed or overly affected by emotions.

24. **TEMPLE VEIN THROBBING**—When a person is under stress, the superficial temporal veins (those nearest the skin on the sides of our heads and just behind the eyes) might pulse or throb visibly. It is a very accurate indicator of autonomic arousal due to anxiety, concern, fear, anger, or, occasionally, excitement. Autonomic arousal is the brain's way of automatically going into survival mode—compelling the heart and the lungs to work faster in anticipation of physical activity such as running or fighting.

25. **FOREHEAD MASSAGING**—We tend to massage our foreheads when we have headaches (literally), when we are processing information, or when we have worries, concerns, doubts, or anxiety. It is a pacifying behavior, which helps to soothe tension or apprehension.

26. **POINTING AT FOREHEAD**—Pointing a finger at the forehead or making a screwing motion with the finger while pointing at the forehead is very insulting—it means that the observer is ill informed, stupid, or crazy. This is a culturally based cue, generally seen in Germany, Switzerland, and Austria, where it is very offensive, and sometimes in the U.S. Because it is insulting it should be avoided.

27. **PRESSING HAND ON FOREHEAD**—Pressing the hand flat against the forehead helps relieve tension caused by stress, doubt, or insecurity. This is different from slapping the hand on the forehead; it looks as though the person is trying to push his head backward. As with so many other behaviors, this is intended to soothe the individual psychologically through tactile pressure on the skin.

28. **PUZZLED LOOK**—The area between the eyes is pulled together, often causing furrowing or knitting of the eyebrows. The eyes may squint or look away, and sometimes the head is canted slightly to the side. We often

see this distressed look when someone is struggling with something mentally or trying to work through a problem. It usually results from a high *cognitive load* (arduous thinking or recalling).

29. COVERING FOREHEAD WITH HAT—Stress or embarrassment will cause some to actually cover their forehead with headgear (a hat, visor, or hood). We generally see this in children and teenagers but also sometimes in adults. I have often watched drivers do this when being ticketed for speeding. It is almost as if they are trying to hide in shame.

THE EYEBROWS

The eyebrows lie just above the supraorbital arches of the eye sockets and serve a variety of purposes. They protect our eyes from dust, light, and moisture, but they also communicate how we feel. From an early age we rely on people's eyebrows to help us interpret their facial expressions. And in many cultures, eyebrows are an aesthetic concern: something to be tweezed, shaped, plucked, colored, highlighted, waxed, stylized, removed, or extenuated. Like the rest of our face, the eyebrows are controlled by a variety of muscles (*corrugator supercilii* principally, but also the *nasalis* and *levator labii superioris* from our nose), and thus can be very expressive and communicate exquisitely our feelings.

30. **EYEBROW ARCHING/FLASHING (HAPPY)**—Eyebrow arching or flashing conveys excitement (such as when greeting a close friend) or the recognition of something pleasing. We arch our brows in less than one-fifth of a second. It is a *gravity-defying behavior*, as it is performed in an upward direction, and as with most

gravity-defying behaviors, it signifies something positive. Babies just a few months old light up when their mother flashes her eyebrows. Here is a great behavior to let others know we care and are happy to see them. A happy eyebrow flash can be immensely useful and powerful in everyday situations both at home and at work.

31. **EYEBROW GREETINGS**—We flash our eyebrows when we recognize someone we know and cannot speak up at that moment, or simply to recognize a person's presence, with or without a smile, depending on circumstances. We are quick to notice when this courtesy is not extended to us, for example, when we enter a store and the clerk makes no effort to establish any kind of eye contact. We can let others know we value them, though we may be occupied, with a very simple eyebrow flash.

32. **EYEBROW ARCHING (TENSE)**—This occurs when a person is presented with an unwanted surprise or shock. Coupled with other behaviors such as a tense face or lip compression, it can let us know someone has experienced something very negative. It is the tension in the muscles that control the eyebrows that differentiates this behavior from the eyebrow greeting described above and it is held for a few seconds longer.

33. **EYEBROW ARCHING (CHIN TOWARD NECK)**—We arch our eyebrows with our mouths closed, chin toward the neck when we hear something we immediately question or are very surprised to hear or learn. When we witness an embarrassing situation we also employ this behavior, as if to say, "I heard that and I didn't like it." It is a look teachers often give to misbehaving students.

34. **EYEBROW ASYMMETRY**—People use this signal when they have doubts or uncertainty. One eyebrow will arch high, while the other remains in the normal position or sinks lower. Asymmetry signals that the person is questioning or doubting what is being said. The actor Jack Nicholson is famous for questioning what others say, on- and offscreen, by this method.

35. **EYEBROW NARROWING/KNITTING**—The area between the eyes and just above the nose is called the *glabella*, and when the glabella becomes narrow or furrowed, it usually means there is an issue, concern, or dislike. This universal sign may happen very quickly and thus can be difficult to detect, but it is an accurate reflection of sentiments. Some people will knit their brow when they hear something troubling or are trying to make sense of what they're being told. The sentiment is communicated with the >< emoji.

THE EYES

Our eyes are the visual gateway to the world around us. From the moment we are born, we are scanning for information in familiar faces, movement or novelty, color, shading, symmetry, and always for the aesthetically pleasing. Our visual cortex, large in proportion to the rest of the brain, seeks novelty and new experiences. Our eyes show love and compassion as well as fear and disdain. Welcoming or joyous eyes can make our day. But eyes can also let us know that something is wrong, that there are worries or concerns. Eyes can own a room or cower in a crowd of strangers. We adorn our eyes to attract and avert them to avoid. They are usually the first thing we notice in others, which is why when a baby is born we spend so much time looking at the eyes. Perhaps because we truly are looking through the window to their soul.

36. PUPIL DILATION—When we are comfortable or like something or someone we encounter, our pupils dilate. We have no control over this. When couples are at ease around each other their pupils dilate as their eyes try to soak up as much light as possible. This is why dimly lit restaurants are a good place to meet, as it naturally softens the eyes and makes the pupils larger—an effect that makes us relax even more around others.

37. **PUPIL CONSTRICTION**—Our pupils constrict when we see something we don't like or when we have negative emotions. Pupil constriction is easier to detect in light-colored eyes. Pupils suddenly shrinking to pinpoints suggest something negative has just transpired. Interestingly, our brain governs this activity to make sure that our eyes are focused in times of distress, as the smaller the aperture, the greater the clarity. This is why squinting improves focus.

38. **RELAXED EYES**—Relaxed eyes signal comfort and confidence. When we are at ease, the muscles around the eyes, the forehead, and the cheeks relax—but the minute we are stressed or something bothers us, they become tense. Babies often demonstrate this quite strikingly, as their facial muscles suddenly scrunch up before they begin to cry. When trying to interpret any body-language behavior, always refer back to the eyes for congruence. If the orbits (eye sockets) look relaxed, chances are all is well. If suddenly there is tension around the eyes or squinting, the person is focusing or might be stressed. The muscles of the eyes and the sur-

rounding tissue react to stressors much more quickly than other facial muscles do, offering almost immediate insight into a person's mental state.

39. **EYE SOCKET NARROWING**—When we feel stressed, upset, threatened, or other negative emotions, the orbits of the eyes will narrow due to the contraction of underlying muscles. The brain immediately makes the eye orbits smaller in response to apprehension, concern, or doubt. It is a good indicator that there is an issue or something is wrong.

40. **QUIVERING UNDER EYES**—The tiny muscles directly under the eyes (the inferior underside of the *Obicularis oculi*) and just above the cheekbones, as well as the surrounding tissue, can be very sensitive to stress. When there is concern, anxiety, or fear, these soft areas will quiver or twitch, revealing the person's true emotional state.

41. **BLINK RATE**—Blink rates can vary depending on environment and the amount of stress or arousal a person is experiencing. Each individual is different, but a typical rate is between sixteen and twenty blinks per minute, depending on lighting conditions and humidity. People looking at computers blink less (many of whom complain of dry eyes or eye infections—tears have antibacterial properties), while those who work

where there is dust or pollen will blink more. Also, be aware that wearing contact lenses can increase how often we blink. When we are around someone who arouses us, our blink rate also tends to increase.

42. **FREQUENT BLINKING**—People who are nervous, tense, or stressed will generally blink more rapidly than those who are not. Frequent blinking is erroneously associated with deception. It is only indicative of stress or other factors noted above as even the honest blink more frequently when being questioned aggressively.

43. **EYE CONTACT**—Eye contact is governed by cultural norms and personal preferences. In some cultures it is permissible to look at someone for three to four seconds, while in others anything beyond two seconds is considered rude. Culture also determines who can look at whom. Even in America, eye contact is determined by what area of the country you are from. In New York City, staring at someone for more than a second and a half might be perceived as an affront. Particular ethnic and cultural groups have their own norms. For instance, many African American and Hispanic children are taught to look down when addressed by elders, as a form of respect.

44. **EYE AVOIDANCE**—We avoid eye contact when it is inconvenient to talk to someone, or when we find a

person unlikable, obnoxious, or repressive. In prison, for example, inmates will avoid eye contact with jailers or inmates known to be aggressive. Eye avoidance can be temporary or long term. Temporarily, people might avert their eyes when a person does something embarrassing. And in the United States, unlike other parts of the world, when we are in close proximity, as in an elevator, we tend to avoid making eye contact with strangers and even with those we know, especially if there are strangers present. Eye avoidance is not indicative of deception, but it can indicate shame or embarrassment.

45. **GAZE SUPERIORITY**—All over the world, studies have shown that high-status individuals engage in more eye contact, while both speaking and listening. Less powerful people tend to make more eye contact with these higher-status individuals while listening but less while speaking. In Japan as well as other Asian Pacific countries this is even more pronounced. Incidentally, we tend to favor individuals who make direct eye contact with us, especially if they are of higher status. Eye contact from high social status individuals, movie stars, for instance, makes us feel favored.

46. **EYE-CONTACT SEEKING**—When we are interested in starting a conversation, whether in a social or a dating

environment, we will actively scan until we make eye contact that says "I am here—please talk to me."

47. **GAZE AND SENTIMENTS**—Around the world, those who study dating cues have noted that oftentimes the first clue that people's feeling for each other have changed is how they look at each other. Long before words are exchanged, the look of increased interest telegraphs that the relationship is changing from friendly to more intimate. How Julie Andrews (as Maria) began to change the way she looked at Christopher Plummer (Captain Von Trapp) in the movie *The Sound of Music* or how Emma Stone (Mia) changed the way she looked at Ryan Gosling's character (Sebastian) in *La La Land* is emblematic of how our gaze changes to reflect our changing sentiment before our words do. It is true in real life as well as in the movies.

48. **GAZE ENGAGING**—This is a behavior intended to get the attention of another person in a warm or romantic way. What makes this behavior stand out is the softness of the face and the repeated attempts to connect, eye to eye, always with a gentleness of the eyes, face, and mouth. We most often see this in dating settings, where it lets the other person know you are interested in further contact or proximity. I have seen strangers engage gazes across broad spaces, communicating their yearning.

49. **GAZING VERSUS STARING**—There is a big difference between gazing at someone and staring at someone. Staring tends to be more impersonal, distant, or confrontational, signaling that we find someone suspicious, alarming, or odd. On the other hand, gazing signals that we take comfort in someone, a much more inviting behavior. When we stare we are on alert; when we gaze we are intrigued, even welcoming. Staring can trigger offense, especially in close quarters such as a bus or subway.

50. **CLOSED EYES**—During a meeting, someone with closed eyes that take a long time to open or that suddenly shut and remain so for longer than usual is probably having issues. It is a *blocking behavior* that reveals dislike, concern, disbelief, or worries—some form of psychological discomfort. Long delays in eye opening reveal deep concern. Conversely, in an intimate setting, closed eyes say, "I trust you, I am blocking everything else out, and I am in the moment with my other senses." Notably, even children born blind will cover their eyes when they hear things they don't like or they find troubling.

51. **EYES CLOSING FOR EMPHASIS**—Oftentimes, when we want to emphasize something or agree in congruence, we will close the eyes ever so briefly. It is a way of affirming what is being said. As with all behaviors,

context is key to ensure it is not a reflection of disagreement.

52. **COVERING OF EYES**—Sudden covering of the eyes with a hand or fingers is a blocking behavior associated with a negative event, such as the revelation of bad news or threatening information. It also indicates negative emotions, worry, or lack of confidence. You also see it with people who have been caught doing something wrong. As I note above, congenitally blind children will also do this, though they cannot explain why; clearly this behavior has an ancient evolutionary basis.

53. **EYES CLOSED, RUBBING BRIDGE OF NOSE**—Individuals who close their eyes and rub the bridge of their nose at the same time are transmitting that they are concerned or worried. This is both a blocking behavior and a pacifier, usually associated with negative emotions, dislike, insecurities, concern, or anxiety.

54. **CRYING**—Crying serves a variety of personal as well as social purposes, most notably providing a cathartic emotional release. Unfortunately, children also learn quickly that crying can be used as a tool to manipulate, and some adults don't hesitate to use it similarly. In observing a person's behavior, crying should not be

given any more weight than other signals that a person is having a hard time. Crying, if it occurs with great frequency, can also let us know when someone is clinically depressed or struggling psychologically.

55. **CRYING WHILE CLUTCHING OBJECTS**—Individuals who cry while clutching at their neck, necklace, or shirt collar are likely undergoing more serious negative emotions than a person merely crying.

56. **EYES DARTING**—Eyes that dart back and forth feverishly are usually associated with the processing of negative information, doubt, anxiety, fear, or concern. Use this behavior in conjunction with other information such as facial tension or chin withdrawal (see #184) to provide a more accurate assessment. It should be noted that some people will dart their eyes back and forth as they analyze a situation, consider options, or think of solutions. This behavior alone is not itself indicative of deception.

57. **EYE-ACCESSING CUES**—As we process a thought, an emotion, or a question posed to us, we tend to look laterally, downward, or up and to the side. This is referred to as *conjugate lateral eye movement* (CLEM) in the scientific literature. There has been a myth for decades, now well debunked by more than twenty studies, that

a person looking away or to the side while answering a question is being deceptive. All we can say when someone looks in a certain direction as they process a question or as they answer it, is that they are thinking—it is not per se indicative of deception.

58. EYELIDS FLUTTERING—Sudden eyelid fluttering suggests that something is wrong or that a person is struggling with something (think of the actor Hugh Grant, who often flutters his eyes on-screen when he has issues or has messed something up). People often flutter their eyes when they are struggling to find the right word or can't believe what they just heard or witnessed. Incredulity is often observed as eyelid fluttering.

59. EYE POINTING—In some cultures an index finger just under an eye communicates doubt or suspicion. But many people across cultures also do this subconsciously in the form of a light scratching motion as they ponder or question something being said. When traveling abroad, ask locals if this means anything special. In Romania, I was told that the finger under the eye was a sign often used to communicate "Be careful, we don't trust everyone who is listening."

60. EYE-POINTING CLUSTER—Pointing of the index finger just under the eye (see #59) clustered with *eyebrow arching* and *compressed lips* simultaneously conveys doubt, bewilderment, or incredulity. This is especially accurate if the chin is tucked in rather than jutted out.

61. **EYE ROLLING**—Rolling of the eyes communicates contempt, disagreement, or dislike. Children often do it to their parents to communicate contention or rebellion. It has no place in a professional setting.

62. **EYELID TOUCHING**—Eyelid touching can be a form of eye blocking coupled with tension relief. Often when people say something they shouldn't have, people nearby will touch or scratch their closed eyelid—this is a good indicator that something improper was uttered. You see this often with politicians when one misspeaks and another catches it.

63. **FATIGUED EYES**—Fatigue usually shows in the eyes first. The eyes and the area around them look strained, puffy, weathered, even discolored. This may be due to long hours working; external factors, such as stress; or crying.

64. **FAR-OFF LOOK**—When alone, or even in conversation with others, staring into the distance, avoiding distractions, allows some people to think or contemplate more effectively. This may be a signal not to interrupt someone when they are deep in thought or recollection.

65. GLAZED EYES—Any number of things can cause the eyes to look glazed, including drugs such as marijuana and alcohol as well as more dangerous substances. When trying to assess whether a person is under the influence of drugs or alcohol, an observer will want to take other behaviors into consideration, such as slurred speech or slowness to respond.

66. LOOKING ASKANCE—Looking askance (sideways) is often used to show a person's doubt, reluctance to commit, disregard, suspiciousness, or even contempt. It is a universal look that reflects disbelief, concerns, or incredulity.

67. LOOKING AT CEILING OR SKY—We often see this dramatic look upward at the sky, with the head tilted back, when suddenly things seem impossible or a person has had a run of bad luck. We see this in sports, such as when a golfer misses a putt. It is a look of disbelief, as if imploring someone on high, in the heavens, to help us or take pity on us. This behavior does have some utility; stress causes tension of the neck, which this position can help relieve by stretching the *sternocleidomastoid* muscles of the neck.

68. LOOKING FOR ACCEPTANCE—When individuals lack confidence or lie, they tend to scrutinize their audience, scanning faces to see if they are being believed. This behavior is not necessarily demonstrative of deception, only of seeking acceptance for what is said. A rule of thumb: the truth teller merely conveys, while the liar often tries to convince.

69. **EYES LOWERED**—This is different from eye avoidance in that the individual does not break eye contact but rather shows deference, piety, humility, or contriteness by slightly lowering the eyes so that eye contact is not direct or intense. This is often culture based, and we see it frequently with children who are taught not to look back at elders or authority figures when being chastised. Black and Latino children are often taught to look down as a form of respect, which should in no circumstances be confused for an attempt to deceive. In Japan it is rude to stare intently at the eyes of a person you meet for the first time; at a minimum, the eyelids must be lowered out of social deference.

70. **SAD EYES**—Eyes look sad, dejected, or depressed when the upper eyelids droop and seem to have no energy. The look may be similar, however, to eyelids drooping from fatigue.

71. **LOOKING AWAY**—Looking away when conversing has to be viewed in context. When there is psycho-

logical comfort, such as when talking to friends, we may feel relaxed enough to look away as we tell a story or remember something from the past. Many individuals find looking away helps them recall details. Looking away is not an indication of deception or lying.

72. **LONG STARE**—In conversations, silence is often accompanied by a long stare. It can be directed at a person or at something in the distance; it merely indicates that the person is in deep thought or processing information.

73. **SQUINTING**—Squinting is an easy way to register displeasure or concern, especially when we hear or see something we don't like. Some people squint whenever they hear something bothersome, making this an accurate reflection of their feelings. But keep in mind that we also squint when we are simply focusing on something or trying to make sense of something we have heard, so context is crucial in interpreting this behavior.

74. **SQUINTING (SLIGHT)**—Often when we are subduing anger we will squint slightly with lowered eyelids. This behavior (narrowing of the slits of the eyes) must be considered in context with other behaviors such as facial tension or, in extreme circumstances, the making of a fist.

75. **STARING AGGRESSIVELY**—A stare can intimidate or serve as the prelude to an altercation. Aggression is signaled by the laser-like focus on the eyes, with no attempt to look away or even blink. Interestingly, other primates also engage in this behavior when observing behaviors that are not tolerated or when there is about to be a physical confrontation.

76. **ANGRY EYES**—Anger is usually displayed by a constellation of facial cues beginning with the distinctive narrowing of the eyes near the nose (like this: > <), coupled with a wrinkled or dilated nose and sometimes the pulling back of the lips to reveal clenched teeth.

77. **EYES WIDENING (STIFF)**—Eyes that remain wide usually indicate stress, surprise, fear, or a significant issue. If the eyes remain stiffly wide longer than usual, something is definitely wrong. This is usually caused by an external stimulus.

78. **EYE ADORNMENT**—Since the time of the Egyptian pyramids, women and men across the globe have adorned their eyes (eyelids, under the eye, the sides, etc.) with a variety of colors to make themselves more aesthetically appealing. Using inks, dyes, minerals, and oils, people have made this part of their cultural traditions, and it has been passed down to our modern

society for a reason: it works. We are attracted to eyes, even more so when they are adorned with colors. We are also attracted to long, thick eyelashes—something that mostly women but some men accentuate to make themselves more appealing.

THE EARS

Cute ears, little ears, sagging ears, deformed ears, big ears, perforated ears, adorned ears. Our ears stick out—sometimes quite literally—and serve some obvious practical functions, from collecting information through sound waves to helping us dissipate heat. But the ears have other utilities you might not have thought about, offering significant nonverbal communication. We know from research that in the early stages of a relationship, lovers spend time studying each other's ears—how they are shaped, how warm they are, how they respond to human touch and even emotions. The ears communicate much more than we think, and in ways that can be quite surprising.

79. **EARLOBE PULLING OR MASSAGING**—Pulling on or massaging the earlobe tends to have a subtle, soothing effect when we are stressed or merely contemplating something. I also associate earlobe rubbing with doubt, hesitation, or weighing of options. In some cultures it means that a person has reservations or is not

sure about what is being said. Actor Humphrey Bogart was notorious for playing with his earlobe as he pondered questions.

80. **EAR FLUSHING OR BLUSHING**—Sudden, noticeable flushing of the skin of the ear, as with other parts of the body (face, neck) may be caused by anger, embarrassment, hormonal changes, reactions to medicine, or autonomic arousal caused by fear or anxiety. The skin covering the ear turns pink, red, or purplish. The skin might also feel hot to the touch. Just having one's personal space violated might cause this reaction. Most people have no control over skin blushing (*hyperemia*) and for some it is very embarrassing.

81. **EAR LEANING**—Turning or leaning our ear toward a speaker conveys that we are listening intently, we want something repeated, or we are hard of hearing. This may be followed by cupping of the ear to literally collect more sound. In dating, we will allow someone we like intimately to draw near our ear, especially when it is extended in that person's direction.

82. **LISTENING**—Active listening is an essential nonverbal in both professional and personal settings. It communicates that we are interested, receptive, or empathetic. Good listeners yield their turn, wait to speak, and are patient when others are speaking. To accom-

plish this we make sure that we face the person we are interested in hearing so that both ears can receive the message.

83. **EAR ORNAMENTATION**—There are any number of ways to decorate, deform, perforate, color, plug, or change the natural look of the ears to fit cultural norms. Ear ornamentation is mostly culture-specific and serves a clear purpose—to communicate social status, courtship availability, or group identification. Ear ornamentation often gives us very accurate insight into a person's background, occupation, social status, heritage, or personality.

84. **SCARRED EARS**—Heat, chemicals, or trauma can damage ear cartilage and tissue. Rugby players, wrestlers, and judokas are susceptible to damaged ears, sometimes called "cauliflower ears."

THE NOSE

At birth, all mammals' noses seek out the mother's milk, which allows them to survive. As humans grow older, our noses continue to help us find the foods we like and to keep us safe, warning us of food that is putrefied or of odors that would do us harm, while helping to filter the air that enters our lungs. When it comes to romance and intimacy, our noses pick up on others' pheromones, making us draw closer while helping us determine subconsciously whether or not we like a person. We may pierce our noses or shape them, as a result of cultural cues, to be thinner, wider, less curved, or more petite. The muscles that cover and surround the nose are so sensitive that when we dislike what we smell, they immediately contract, wrinkling our noses to reveal our disgust. Noses help to distinguish us from others physically, they protect us from harmful chemicals and bacteria, and as you will see, they are essential to communication and to understanding others.

85. **COVERING NOSE WITH BOTH HANDS**—The sudden covering of the nose and mouth with both hands is

associated with shock, surprise, insecurity, fear, doubt, or apprehension. We witness this at tragic events such as car accidents and natural disasters as well as when someone receives horrible news. Evolutionary psychologists speculate that this behavior may have been adapted so that predators, such as lions or hyenas, would not hear us breathing. It is seen universally.

86. **NOSE WRINKLING UPWARD (DISGUST)**—The signal or cue for disgust usually involves the nose wrinkling upward (also known as a "bunny nose"), while the skin contracts along with the underlying muscle (the *nasalis*), which is very sensitive to negative emotions. Often this gesture will cause the corners of the eyes near the nose to also narrow. Babies, beginning at the age of roughly three months and sometimes even earlier, will wrinkle their noses when they smell things they don't like. This disgust cue remains with us all our lives. When we smell, hear, or even just see something we don't like, our nasalis muscle contracts involuntarily, revealing our true sentiments.

87. **UNILATERAL NOSE WRINKLING**—As noted above, nose wrinkling or crinkling upward is an accurate indicator of dislike or displeasure and usually occurs on both sides of the nose. However, there are people in whom this occurs only on one side of the nose (uni-

laterally). As the nose muscles pull upward, wrinkling just one side, they also tend to pull the upper lip of that side of the face. Some people call it the Elvis effect. When the side of the nose is noticeably pulled up, it means the same thing as the full nose wrinkle—dislike.

88. **NOSE TWITCHING (CARIBBEAN)**—This behavior is somewhat similar to the disgust display above (see #86) but occurs much faster, sometimes in as little as 1/25th of a second. When a person looks directly at someone, the nose muscle will contract rapidly, wrinkling the nose upward—but without the eyes squinting as in the disgust cue above. This behavior is a linguistic shortcut that wordlessly asks "What's going on?" "What happened?" "What do you need?" It is seen throughout the Caribbean, including in Cuba, Puerto Rico, and the Dominican Republic, and thus also found in U.S. cities that have large Caribbean populations such as Miami and New York. At the Miami International Airport, I'm frequently greeted at the coffee counter with this nose twitch, which means "What can I get you?" If you see it, just place your order.

89. **INDEX FINGER TO NOSE**—Placing the index finger under the nose or on the side of the nose for a period of time is sometimes associated with pensiveness or

concern. Look for other clues to help you discern what it means. This behavior is different from sneaking a nose feel (see #95) or nose stroking, as in this case the finger just lingers there for a long time.

90. NOSE BRUSHING—This distinctive behavior of brushing one's nose very lightly several times with the index finger is usually associated with stress or psychological discomfort, though it can also present in someone pondering something dubious or questionable.

91. HOLDING NOSE HIGH—A high nose profile—an intentional tilting of the head, with the nose pointed upward—indicates confidence, superiority, arrogance, or even indignation. It is a cultural display, seen in some countries and societies more than in others. It may signal superiority, such as when high-status individuals affirm their rank at the start of a meeting. Italian dictator Mussolini was famous for this, as was General Charles de Gaulle of France. In Russia, the ceremonial guards at the Kremlin are notorious for this nose-high behavior.

92. NOSE TAPPING/SIGNALING—In many cultures a very overt tapping of the nose with the index finger can mean "This stinks," "I don't trust you," "I question this," or "I am watching you very carefully." It can also mean "I notice you," "You are very clever," or "I acknowledge you" (Paul Newman and Robert Redford did this to each other in the movie *The Sting*).

93. **NOSTRIL FLARING**—We usually flare our nostrils (*naral wings*) in preparation for doing something physical. Frequently, people who are upset, feel they have to get up or run out, or are about to violently act out will flare their nostrils as they oxygenate. In police work it may signal a person is about to run. Interpersonally, it is a good marker that a person needs a moment to calm down.

94. **PLAYING WITH PHILTRUM**—The grooved area just above the upper lip and below the nose is the *philtrum*. People will play with this area by plucking at it, scratching it, or pulling on it when stressed—sometimes rather energetically. The philtrum is also revealing in other ways—sweat tends to gather there when people are stressed. They might also place the tongue between the teeth and the back of the philtrum, pushing it out. Stimulation of this area with the tongue is an easily spotted pacifier.

95. **SNEAKING NOSE TOUCH**—Sneaking a pacifying touch by ever so slightly rubbing the nose with the index finger indicates tension that is being masked and the need to convey the perception that everything is fine. Look for it from professionals who are accustomed to being in

control but are under stress. It is also often seen in poker players who are trying to hide a weak hand.

96. **RAPID NOSE INHALING**—Many people, when about to deliver bad or unpleasant news, will rapidly inhale through the nose, loudly enough to be heard, before they speak. I have also seen people do this as they hear a question that bothers them, and in some instances before they lie. The hairs and the nerves in the nose are very sensitive to moisture as well as air movement and touch. The quick inhale stimulates the hairs and the connected nerve endings, which appears to momentarily mitigate the stress of having to say or reveal something that is troubling.

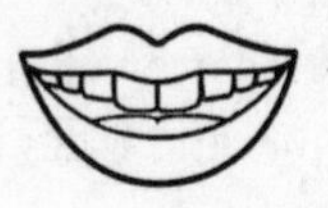

THE MOUTH

The mouth is essential for eating, breathing, and drinking and is also, of course, where we form and pronounce words. Highly sensitive to touch and temperature, the mouth is surrounded by more than ten intricately reflexive muscles that not only respond to touch but also reflect our thoughts and sentiments. The mouth can be seductive or sad, joyous or pained—and it accurately registers when one emotion gives way to another in an instant. After we look at the eyes for information, it is here where we search for additional cues as to what is in the mind.

97. **LOUD, SHORT EXHALING**—This type of exhale, where the lips are left slightly open, indicates high stress or frustration. People exhibit this behavior when hearing bad news or when confronted with a difficult situation. It helps relieve stress, especially when we are angry.

98. **CATHARTIC EXHALING**—Exhaling with puffed-out cheeks and tight lips indicates that stress is being ex-

perienced or has passed. You might see this when a test or an interview is over or after a near accident. This exhale is very audible and takes longer to perform than the above version.

99. **AFFIRMATIVE INHALING**—A sudden loud inhaling makes a distinct sound that is used in Scandinavian countries, parts of the United Kingdom, and Ireland to signify "Yes" or "Yes, I agree." It is a linguistic shortcut, as no words need to be used. The person quickly inhales loudly enough to sound as if she is gasping for air. Once, after a car ride in Sweden, when I asked if we had arrived, the driver merely did an affirmative inhale—and that was it.

100. **SUCKING IN AIR THROUGH CORNERS OF MOUTH**—This behavior is both seen and heard. The corners of mouth suddenly open slightly and air is quickly inhaled, making a sucking sound. It is extremely reliable in what it reveals: fright, concern, or anxiety. That the majority of the mouth is closed signifies that the person is, in essence, restricting free movement of the lips, an action that suggests stress and in some cases pain, such as when someone steps on your toes.

101. **HOLDING THE BREATH**—Polygraphers know this well: when stressed, many people have an impulse

to hold their breath to try to contain their nervous breathing. Often they even have to be told to breathe. Holding one's breath is part of the *freeze, flight, fight* response. If you see someone restraining their breathing or actually holding their breath when asked a question, most likely they are experiencing fear or apprehension.

102. DRY MOUTH—Stress, fear, and apprehension can cause our mouths to dry out (the clinical term for this is *xerostomia*). Some prescribed medicines as well as illicit drugs may also cause dryness of the mouth. A dry mouth is not, as some believe, indicative of deception. It can, however, indicate that someone is stressed or anxious.

103. SALIVA BALLS OF MOUTH—A dry mouth due to stress, medication, or illness can cause saliva to become dry and clumpy; these clumps—they often look like little cotton balls—tend to collect in the corners of the mouth. They are sometimes noticeable in speakers who are nervous. It is quite distracting. If you are nervous, it is a good habit to pinch and wipe the corners of your mouth to avoid saliva balls as well as drink water. The clinical term for a dry mouth is xerostomia.

104. CHEWING GUM—Gum chewing is an effective pacifier. Chewing vigorously might signal concern or anxiety. Some people, when stressed, will chew rapidly out of habit even if they don't have gum in their mouths.

105. **VOCAL TICS**—Sudden vocal projection of noises, clicks, chirps, or throat clearing can be alarming if one is not acquainted with Tourette's syndrome (TS), or other disorders that contribute to vocal tics. Stress and anxiety may be the catalysts for Tourette's outbursts, and there is nothing for us to do but recognize that this is out of the person's control. It is also not uncommon to see the arms move erratically. The best we can do is encourage others not to stare, as this is embarrassing for the person with TS.

106. **TONGUE BITING/CHEWING**—Some individuals under stress will bite their tongue or the inside of their cheeks in order to soothe their nerves. It is very pronounced in those for whom it has become a nervous tic. The tongue will appear wounded or even ulcerated in places. Under stress the behavior is of course heightened. Unfortunately, tongue and cheek biting, like repeated hair pulling, can become pathological.

107. **MOUTH STRETCHING**—When we are afraid or realize we made a mistake, we often find ourselves involuntarily exposing the bottom row of clenched teeth as the corners of the mouth stretch substan-

tially downward and to the side. This is often seen when we are reminded that we forgot to bring something important.

108. YAWNING—Yawning is an excellent pacifier, as it relieves pent-up stress by stimulating nerves in the jaw; specifically the *temporomandibular* joint. It was also recently discovered that the rapid intake of air when we yawn cools the blood circulating within the palate of the mouth and, like a car radiator, the blood going to the brain. Yawning may indicate that someone is too hot or, as I often found during interviews, that an interviewee was severely stressed. Babies wrapped too warmly will also yawn with greater frequency as they sleep to help them cool down.

109. SMOKING—People who smoke do so more often when they are stressed. Note any deviations from a person's normal smoking routine as evidence of how stressed they may be. They may be so stressed they lose count of how many cigarettes they have lit. Excessive smoking also leads to tobacco stains on the fingers and, of course, the stench in their hands.

110. OVEREATING—Under stress some people will overeat, sometimes going far beyond their normal food intake. I have seen people during a football game consume vast amounts of food, to the point of getting sick, their anxiety over the status of their favorite team transferred to their appetite.

111. TONGUE IN CHEEK—Pushing the tongue firmly against one cheek and holding it in place serves to relieve tension. This is most often seen in individuals facing high stress or in those hiding information or who are getting away with something. It can also be seen in those who are being playful or cheeky.

112. TONGUE JUTTING—When the tongue suddenly protrudes between the teeth, sometimes without touching the lips, it means "I got away with something" or "Oops, I got caught." You also see it when people catch themselves making a mistake. The tongue jut is universal and is remarkable in its consistency, whether it is indicating that you got away with a great bargain or an extra cookie, a higher grade, or a whopper of a lie.

113. TONGUE INSULTS—In almost all cultures the sticking out of the tongue is used as an insult, a display of disgust or dislike. Children use this technique from a very young age when they want to insult one another. Pacific Island warriors such as the Māori will dramatically stick their tongue out and down as a way to intimidate and insult. Coupled with very wide eyes, a stuck-out tongue can be quite intimidating, and it is still used to this day in Māori *haka* ceremonies.

114. **TONGUE PROTRUDING**—Oftentimes, while performing a complex task, people will stick out their tongue, usually to one side or the other, or drape it over their lower lip. I had an accountant who did this as he entered numbers into a calculator, and I see it all the time at the university when students are taking tests. This tongue placement serves dual purposes: it pacifies us while simultaneously communicating to others that we are busy and should not be disturbed. Michael Jordan famously did this while playing basketball; when his tongue was out, two points usually soon followed.

115. **TONGUE PRESSING AGAINST PALATE**—People might press their tongue against the roof of their mouth when they are struggling with something. It is seen in people taking tests, filling out applications, after missing a shot in basketball, or when somebody needs psychological comforting. The mouth is generally left slightly open, allowing observers to at least partially see the tongue.

116. **TONGUE LICKING TEETH**—As with lip licking (see #145), we lick our teeth when our mouth is dry—usually due to nervousness, anxiety, or fear. The rubbing of the tongue across the teeth and/or gums is a universal stress reliever, as well as a potential sign of dehydration. Incidentally, when this is done with the mouth closed, you can see the tongue track across the teeth under the lips.

117. **TONGUE DARTING**—To relieve stress some people will dart their tongue back and forth from corner to corner of their mouth (noticeable through the cheeks) in nervous or worried anticipation. Usually they think they are not being noticed or that the meaning of this behavior cannot be deciphered.

118. **FLICKING NAILS ON TEETH**—The flicking of the thumbnail on the teeth releases stress. People who do this repeatedly are trying to soothe themselves because they are anxious about something. Keep in mind, however, that as with all repetitive behaviors, if people do it all the time, then you ignore that behavior because that is their "norm"—it may be more significant when they stop doing it.

119. **TEETH BARING**—Sometimes people suddenly pull the corners of the mouth back and hold that position while they show their clenched teeth. This is a legacy "fear grin" very similar to what chimpanzees do when they are scared or fear a dominant male. We humans tend to bare our teeth this way when we get caught doing something we shouldn't be doing. This behavior might be coupled with a simultaneous arching of the eyebrows, depending on the circumstances.

120. **TEETH TAPPING**—When stressed, bored, or frustrated, some people will shift their jaw slightly and tap

their canines together, favoring one side of the mouth or the other. This sends repetitive signals to the brain that help soothe us.

121. **VOICE TONE**—The tone of our voice can make people comfortable or feel like we are challenging them. We can use the tone of our voice to alter or enhance how we are perceived. You can come off as nice, sweet, kind, loving, and knowledgeable, depending on your tone of voice or alternatively as suspicious, indignant, or arrogant. Tone of voice matters greatly. Ironically, if you want to get people's attention, lowering your tone of voice will work best. A lower voice is also soothing, as any parent who has put a child to bed will attest.

122. **VOICE PITCH**—When we are nervous our voices tend to rise in pitch. Listen for voices that rise or crack when a person is stressed, nervous, or insecure. This is caused by vocal-cord tension.

123. **UPTALK**—Uptalk is when people inflect their tone up at the end of a declarative sentence, as though it were a question. Studies show that even a single instance of uptalk on the phone can negatively impact the listener's impression of the speaker. Though uptalk is popular with many young people, it makes them sound tentative and lacking in confidence.

124. **STUTTERING/STAMMERING**—Some individuals pathologically stutter (repeating syllables as they try to speak). For some it can be quite debilitating, as in the case of England's King George VI, famously depicted by Colin Firth in the 2010 movie *The King's Speech*. For many of us who do not stutter pathologically, a high degree of stress or anxiety can cause us to temporarily stutter or stammer.

125. **DELAY IN ANSWERING**—Many people erroneously believe that a delay in answering a question signals that a person is lying or is buying time in an attempt to muster a credible answer. Unfortunately, both the honest and the dishonest may delay an answer but for different reasons. The guilty may in fact have to think about what to say while the innocent may be thinking about how best to say it. In my experience, a delay in answering should make us take note but is not indicative of deception. In some cultures—for instance, among many Native Americans—a delay in answering is not unusual as the person contemplates the complexity and nuance of a question. Stress or fatigue can also make us slow to answer. A formal inquiry may also cause us to delay answering because of the seriousness of the hearing.

126. **SILENCE**—A prolonged silence, or even just a "pregnant pause," may speak volumes. Sometimes, when we cannot remember information or we are contemplating

something, a silence is unintentional. But other times it is very much intended, as when a negotiator may go temporarily silent to get the other party to fill in the void. Silence can be used to communicate that the person is pondering, recollecting, considering, processing, or is nonplussed. Great actors use it effectively, as do interviewers.

127. **SILENCE AND FREEZE RESPONSE**—When a person suddenly goes silent and stops moving or undergoes breathing changes upon hearing or seeing something, take note. This is a response to something negative that shocks them or causes them to reassess what they know or believe.

128. **INTERRUPTIVE ARGUMENTS**—Arguing for the sole purpose of disrupting a meeting or a conversation is an often-used technique to prevent further discussion. It is the repetitive interruption, not the words used, that is the nonverbal here that distracts or antagonizes. The technique does not further a conversation or provide any clarity, it is clearly intended to aggravate, intimidate, or place someone on emotional "tilt." I have seen this many times in union meetings as members disrupt a speaker.

129. **CATHARTIC UTTERANCES**—In this form of a cathartic exhale, we come close to saying a word but never get there. "Ohhhh" or "woooo" or "fuuuuh" is uttered

but never completed. These are considered nonverbals because the actual words are not spoken, though we can often intuit their meaning. Often these utterances don't make sense, especially to foreigners, but they help us to relieve stress without offending anyone.

130. **SPEED OF TALKING**—How fast we speak is a key nonverbal indicator. In some parts of America people speak very slowly and deliberately, while in others speech is fast and clipped. These styles communicate something about the personality of the speakers—where they are from, where they went to school, and more. Changes in a person's normal speed of talking may indicate stress or reluctance to answer a sensitive question.

131. **INCESSANT TALKING**—We have all met people who seem to never stop talking. They might simply be nervous, or they might be inconsiderate of others and focused only on themselves. Context is key. In the aftermath of an accident, a person might ramble, talking nonstop. This is caused by shock. But at a party, the man who talks your ear off is letting you know who he thinks it most important—and it's not you.

132. **INCONGRUENT TALK**—After an accident or tragic event, a person may begin to speak incoherently. This is a result of stress and the emotional side of the brain

being overwhelmed. Depending on the circumstances of the event or tragedy, this may last for hours or even days, as we have seen with soldiers and refugees in combat zones.

133. **REPETITION OF WORDS**—Under high stress, people may repeat certain words over and over in a nonsensical way. Efforts on your part to get them to say more may not work. It is as if they are stuck in a loop. I once heard a victim struck by a vehicle say the word "metal" over and over again, with a look of fright upon her face. That was all she could say.

134. **SPEED OF RESPONSE**—Some people will take their time answering a question, starting, then stopping, then continuing. Others will respond before you finish asking the question. How fast they answer says something about how they are thinking and processing information. Keep in mind that speed of response depends upon cultural context as well as mental agility.

135. **SPEEDING THROUGH COMMENTS**—Fast is not always good when answering a question. When a person speeds through an apology, the apology loses its meaning—it seems mechanical and contrived. A similar principle applies in praising or welcoming people. It is at these moments that we should take our time. Speeding through an apology or recognition of another

suggests there are issues, such as social anxiety, reluctance, or lack of conviction. It is the speed of talking that is the nonverbal here—as if glancing over what is important.

136. FILLER SOUNDS—Sounds such as "aah," "hum," "hum," coughing or throat clearing, and hesitations in speaking may indicate people are momentarily at a loss for words and feeling they have to fill the void with at least a sound. Americans are notorious for using filler sounds as they figure out what to say, struggle to find the right words, or bide their time while they recall an experience. Because these are not actual words, they are considered a paralanguage or a nonverbal.

137. COUGHING OR CLEARING OF THROAT—People often cough or clear their throat when they need to answer or deal with something difficult. A question that is challenging to answer or needs to be qualified might cause throat clearing. I have noted that some individuals when lying will clear their throat or cough, but this is not a reliable indicator of deception, as the honest may also do so when nervous or tense.

138. WHISTLING NERVOUSLY—Whistling is a form of *cathartic exhaling* (see #98), and it helps us relieve stress. It's a good pacifier and that's why people tend to do it when traveling by themselves through a dark or desolate area or when they feel uncomfortably alone. In movies and cartoons, people or characters are often portrayed whistling while walking through a cemetery to ward off their apprehension.

139. TUT-TUTTING—These tongue and teeth noises are used in many societies to indicate disagreement, to call attention to something that is wrong, or to shame. One tut-tuts by placing the tongue against the back of the front teeth and the upper palate and then rapidly inhaling to make a sharp, quick sound. This is often seen in concert with a waving finger indicating that a transgression has occurred and been noticed. Parents frequently tut-tut when children are about to misbehave.

140. LAUGHTER—Laughter is a universal display of amusement, happiness, and joy. We know that when we laugh we experience less stress and even less pain; indeed, the act of laughing may have arisen in us as a protective evolutionary benefit. There are, of course, many different sorts of laughter: unrestrained cackles when we hear a genuinely hilarious joke; the joyous laughter of children; the obsequious laughter of those who seek to flatter a leader. How someone laughs says a lot, and should be examined for the true depth of sentiment and context when you're in doubt.

THE LIPS

We purse them in front of smartphones to take selfies and paint them with lipstick to make them more attractive. We inject them with collagen to hide our age, and we lick them to keep them moist. Rich in nerve endings, our lips sense pressure, heat, cold, flavors, tenderness, and even the movement of air. They not only sense, they can be sensuous as well. Lips communicate moods, likes, dislikes, even fear. We adorn them, massage them, Botox them, and play with them—and oh yes, we kiss with them. In a way, they are one of the things that make us uniquely human.

141. **LIP FULLNESS**—Our lips change size and dimensions according to our emotional state. They get small when we're stressed, larger when we're comfortable. Full, pliable lips indicate relaxation and contentment. When we're under stress, blood flows out of the lips to other parts of the body where it is needed. Lip fullness can serve as a barometer of a person's emotional state.

142. **FINGERTIPS TO LIPS**—Covering one's lips with one's fingers can indicate insecurity or doubt and should be considered in context. Watch for this behavior, especially as people hear a question they need to process. This behavior is also seen when people carefully ponder an issue. Keep in mind that some people do this frequently, in all sorts of situations—it is a stress reliever harking back to when they sucked their thumbs, so be careful with what inference is drawn.

143. **LIP PLUCKING**—Pulling or plucking of the lips is usually associated with fear, doubt, concern, lack of confidence, or other difficulties. Ignore people who do this continually to pass the time—for them it is a pacifier. For those who rarely do it, it's a good indicator that something is wrong.

144. **LIP BITING**—Lip biting is a pacifier, usually seen when people are under stress or have concerns. We bite our lips because, after a certain age, it is no longer socially acceptable to suck our thumbs, and biting our lips stimulates the same nerves in the mouth. We might also bite our lips when we want to say something but can't or shouldn't. Note also that some people, when angry, will bite their lips as a means of self-restraint.

145. **LIP LICKING**—Rubbing the tongue on the lips helps to pacify us in the same way that lip biting does. This

behavior is usually associated with concerns, anxiety, or negative emotions; however, it could just be that the person has dry lips, so be careful when drawing conclusions. For some people, however, this is a very reliable indicator that they are very stressed. As an educator, I see this all the time when an unprepared student sits down for a test.

146. **LIP NARROWING**—The narrowing of the lips is mostly associated with negative thoughts, concerns, fears, anxiety, or lack of confidence. As we process issues or experience stress, the lips tend to narrow.

147. **LIP COMPRESSING**—Throughout the day, as we encounter negative events or uncomfortable thoughts, and concerns, our lips will narrow and press together, accurately transmitting, even if only for an instant, our concerns. Lip compression can be very subtle or can reach a point where the lips noticeably change color as blood is forced out. Lip compression can be very fleeting (1⁄20th of a second), and yet it reveals accurately a negative emotion suddenly registered.

148. **SLIGHT PRESSING OF LIPS**—Sometimes we show our annoyance with others by slightly compressing the lips. Unlike full lip compression, where both lips are involved, this usually involves only the upper lip. Still, a slight lip compression might reveal something, when

considered along with the rest of a person's body language.

149. **COMPRESSED LIPS PULLED DOWN**—You'll see this striking behavior in people when they realize they made a major mistake or get caught doing something wrong. The lips are held tightly together while the muscles surrounding the mouth contract to bring the lips slightly down, stretching the upper lip away from the nose and pulling the mouth area tightly against the teeth.

150. **RELUCTANCE TO DECOMPRESS LIPS**—People who hold their compressed lips together for a long time, reluctant to decompress them, are signaling a high degree of stress or concern. Lip compression is, in a way, a battening down of our hatches, much like covering our eyes with our hands to block out something negative. The greater the tension or apprehension, the greater the need to keep the lips compressed.

151. **LIP WITHDRAWING**—When we have deep concerns or anxiety, we might suck our lips into our mouth to the point where they are no longer visible. This signals something very different from lip compressing (see #147), where much of the lips remain visible. This behavior is often reserved for when there is severe stress, significant physical pain, or great emotional turmoil.

152. **LIP QUIVERING**—The quivering of the edges of the lips, no matter how slight, in the absence of alcohol or neurological disorders, indicates discomfort, concern, fear, or other issues. Young people when questioned by parents or other adults in positions of authority often display quivering lips, as do honest people who have never been confronted by law enforcement officers before. I have also heard from human resources personnel that some young people's lips will quiver when they are asked if they use illicit drugs.

153. **UPSIDE-DOWN LIPS**—When the lips are compressed and the corners of the mouth turn downward, things are really bad emotionally. This is a strong indicator of high stress or discomfort. This behavior is difficult to fake, so it is very accurate. Be careful, however, because some people have naturally downturned mouths. This indicator is similar to the "grouper" mouth (see #156), but in this case the lips either are very tightly compressed or have disappeared completely.

154. **LIP PURSING**—We purse our lips (pinching them tightly toward the front of the mouth) when we disagree with something or when we are thinking of an alternative. When audiences take issue with what a speaker is saying or know it is wrong, you often see this behavior. The more outward the movement of the pursed lips, the stronger the negative emotion or

sentiment. This is an extremely reliable behavior you also see in poker when players don't like their own hole cards.

155. **LIP PURSING PULLED TO SIDE**—This is similar to the pursed-lips behavior above, but with the lips energetically pulled to the side of the face, significantly altering the look of the person. Usually this happens quickly, though when there is strong disagreement, the position might be held for a few seconds. It is an emphatic gesture that says, "I have real issues here; I don't like what I was asked, what I just heard, or where this is going." The more pronounced the gesture or the longer it is held, the stronger the sentiment. We famously saw this expression on O. J. Simpson trial witness Kato Kaelin as he testified, and gymnast McKayla Maroney when she came in second place in the vault finals during the 2012 Summer Olympics.

156. **SAD MOUTH**—The mouth, like the eyes, can be a window into our emotional state. Sadness is usually shown with the corners of the lips turned down slightly, usually in concert with lowered upper eyelids. This is sometimes referred to as a "grouper" mouth or face. It should be noted that some people naturally look this way—the corners of their mouths perpetually turned down—and for them, it has nothing to do with negative emotions.

157. **THE O**—When we are surprised or in agony, our lips will often instinctively make an oval shape, similar to an O. The reason we do this is not exactly known, but it seems to be a universal behavior across cultures and possibly a vestigial response we share with alarmed primates. The best-known image of this is Edvard Munch's painting *The Scream*.

158. **MOUTH OPEN, JAW TO SIDE**—Similar to jaw dropping (see #179), this occurs when people have done something wrong or realize they've made a mistake. One corner of the mouth is pulled to the side, causing the jaw to shift in that direction; at the same time, the clenched lower teeth on that side of the mouth are exposed. Students often react this way when they miss a question they know they should have known; it's also seen when employees recognize they failed to complete a task. This behavior might be accompanied by the quick sucking in of air through clenched teeth.

159. **SMILE**—A genuine smile is an instant, surefire way to communicate friendliness and goodwill. Around the world it signals warmth, friendliness, and social harmony. Watching someone smile, especially babies, brings us joy. In family relations, dating, and business a smile opens doors as well as hearts. There are a variety of smiles, including social smiles for those whom we don't know but acknowledge near us, the tense smile

of a test taker, and the false smile of those pretending to like us or trying to act comfortable.

160. **TRUE SMILE**—A topic of much research; a genuine smile involves the mouth and the muscles around the eyes. This is called a *Duchenne smile,* according to body-language researcher Paul Ekman. The face is visibly more relaxed in a true smile, as the facial muscles reflect actual joy rather than tension. Studies have shown that a genuine smile can be truly "contagious," in both professional and personal environments, and is often a trait we associate with charismatic individuals.

161. **FALSE SMILE**—False smiles, like nervous smiles, are used for perception management to make others believe everything is OK. They are fairly easy to distinguish from a true smile; however, in a false smile, sometimes only one side of the face is involved, or the smile goes toward the ear rather than the eyes. It looks contrived. A true smile engages the eyes and the facial muscles smoothly on both sides of the face.

162. **NERVOUS SMILE**—A nervous or tense smile shows anxiety, concern, or stress. The nervous smile is performed to make others think everything is fine. You often see this on visitors clearing customs at the airport; they nervously smile at the inquisitive officer asking questions.

163. SMILING AS A BAROMETER OF EMOTIONS—How accurate are smiles in revealing our inner feelings? Very. Studies show that athletes' smiles differ noticeably depending on whether they finish in first, second, or third place. Interestingly, this same distinction holds true for congenitally blind athletes, who have never actually seen a smile on another person's face. Their smile will reflect their success, or lack thereof—again confirming that many nonverbals are hardwired in our brains.

164. CRIMPING CORNERS OF THE MOUTH—When one corner of the mouth pinches tight and pulls slightly to the side or up, it reveals smugness, disdain, dislike, disbelief, or contempt. Where the contempt is demonstrably overt, this behavior may be dramatized or exaggerated, leaving no question as to true sentiments. Most of the time crimping the corner of the mouth is done on just one side of the face but some people do it on both sides and it means the same.

165. UPPER LIP RISE—Disgust, negative sentiments, disdain, or dislike will cause the upper corner of the lip on one side of the mouth to rise slightly or "tent" upward. When the sentiments are strong, the rise can be very noticeable, distorting the upper lip toward the nose and exposing the teeth, almost in a snarl. This is a sign of utter dislike or disgust.

166. UPPER-LIP TONGUE RUBBING—Some people reflect their positive emotions by licking their upper lip briskly back and forth. Because the tongue is in essence defying gravity (going for the upper lip), positive emotions are more likely involved. This is differentiated from the usual lip licking, which is done on the lower lip and is associated with stress release. As with all body-language indicators, there are exceptions, and some people rub the upper lip to relieve stress, so look for other confirming behaviors to guide you in drawing conclusions.

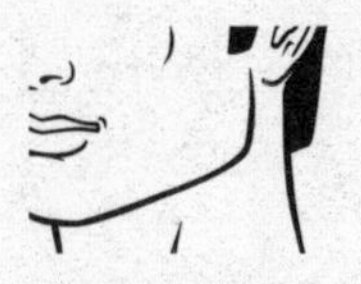

THE CHEEKS AND JAW

Many people think of the cheeks as a dormant fixture and the jaw as something only useful for chewing and talking—not, in other words, useful in the study of body language. But our cheeks and jaws give our faces our unique human shape. We look for leaders to have strong jaws and the fashion industry is always looking for high cheekbones on models. We color our cheeks artificially with makeup to increase our attractiveness and allow hair to grow on our jaws to fill out a face—which is why President Lincoln grew out his beard. From cheeks that flush with excitement or embarrassment to jaws that shift when we feel unsure, these two areas definitely communicate something about us and should not be overlooked.

167. **SUDDEN FACIAL TICS**—Facial tics can erupt anywhere on the face (the cheek, a corner of the mouth, the eyes, the forehead) and are specific to each individual. If you suddenly see a nervous twitch, it is usually caused by tension or anxiety. Facial tics often occur

on or near the cheeks because of the interconnecting muscles that traverse this area.

168. **FACIAL DENTING**—People will push or press their fingers firmly against a cheek to produce sensations that relieve stress—literally making a dent into their own skin. These displays are sometimes quite pronounced, depending on the pressure applied. This is frequently seen at sporting events when the home team is doing poorly. Facial denting can be done with one or two hands or a few fingers on just one side or by pinching the cheeks between the thumb and the index or middle finger in concert.

169. **CHEEK OR FACIAL MASSAGING**—Cheek or facial massaging is a good way to release stress. Usually done very softly, it can also signal contemplation. This is a behavior that needs to be considered with other behaviors for an accurate assessment.

170. **CHEEK STRUMMING**—Strumming the fingers on the cheek indicates that someone is bored and wanting to move things along. Verify with other behaviors, such as looking bored or seat shifting.

171. **CHEEK FRAMING**—Cheek framing is when a person rests the jaw on an extended thumb and places the index finger up along the side of the cheek. This usu-

ally involves just one hand and suggests that a person is pondering something, or wants to appear pensive. Some people use this behavior primarily when they doubt what a speaker is saying, while others might simply do it as a means to aid concentration. In dating, it can be an effective pose to show interest from a distance.

172. **PUFFING OUT CHEEKS**—The puffing out of the cheeks, without exhaling, often signifies doubt, deliberation, or caution. This is often seen in people who are not quite sure what to do next or who are apprehensive. It is not unusual to see someone hold this pose for quite a while as they work out the solution to a problem.

173. **SNEAKING A CHEEK TOUCH**—Sneaking a pacifier by ever so slightly rubbing the index finger against the cheek indicates that stress is being managed for the sake of perception. When people try to conceal a pacifier, like touching the side of the nose, they do so because they are trying to hide their insecurity, anxiety, or worry. Surreptitious cheek touching is frequently noticeable in people being interviewed on TV and in poker players.

174. **CHEEK SCRATCHING**—Cheek scratching is also a pacifier, a way of dealing with doubts and insecurities. It is more robust than sneaking a touch, which

tends to be more accurate because of its hidden meaning. Nevertheless, the scratching of the cheek with four fingers usually indicates reservations, hesitation, bewilderment, or apprehension.

175. PINCHING THE CORNERS OF THE MOUTH—Using the fingers to tightly constrict or pinch the corners of the mouth relieves stress. We rarely do this when we are content and relaxed. It is different from facial denting (see #168). This behavior is usually done by pressing the fleshy area of cheeks with the fingers and thumbs bilaterally pulling toward the corners of the mouth, perhaps even pulling on one or both lips.

176. CHEEK WIPING—Under extreme stress, it is not unusual to see people press their hands on their face and drag them downward, as if wiping their faces clean. Typically, the motion starts just in front of the ears and concludes near the jawbone. The harder and longer the person presses down, the more acute the stress. I've seen stockbrokers do this at the closing bell after a poor day of trading or when a team loses in the final second of a game.

177. JAW TENSING—When we are upset, angry, or fearful, the jaw muscles near the ears tend to tense up. Look for jaw tension when there is stress, defiance, or emotions are becoming heated.

178. JAW DISPLACING/SHIFTING—Jaw displacement or repetitive jaw shifting (from side to side) is an effective pacifier. This is also simply a compulsive behavior in some people, so note when and how often it occurs and look for other confirming behaviors that something is amiss. Most people do this infrequently, and thus when you do see it, it is very accurate in communicating that something is bothering them.

179. JAW DROPPING—A sudden drop of the jaw, leaving the mouth open and the teeth exposed, communicates great surprise. This behavior is often seen when people are shocked or are confronted with an embarrassing revelation. Why our jaws drop is not completely understood, but the action is quite accurate in revealing total surprise.

180. JAW MUSCLES PULSING—Jaw muscles that pulse, throb, or become tight and pronounced indicate impatience, tension, concern, worries, anger, or negative emotions.

181. JAW JUTTING—When we are angry, we tend to move or jut the jaw slightly forward. In conjunction with lowered upper eyelids or tense lips, this behavior makes anger difficult for a person to hide entirely.

THE CHIN

Baby, round, squared, sagging, strong, dimpled, cute, or scarred: chins come in many varieties and shapes. They protect our face, and if need be our neck, but they also communicate our sentiments, whether pride or shame. We say "chin up" when others are down, and soldiers proudly salute the flag with their chins angled high. The chin, in short, can speak volumes about our internal state, whether we are confident, frightened, troubled, or emotionally overcome.

182. **CHIN UP**—When the chin is out and up it communicates confidence—thus the saying "chin up." In certain European cultures (German, French, Russian, and Italian, among others) the chin is generally raised higher than normal to signify confidence, pride, and in certain cases, arrogance.

183. **CHIN POINTING DOWN**—If the chin suddenly points downward in response to a question, most likely the

person lacks confidence or feels threatened. In some people, this is a highly reliable tell; they literally drop their chin when they get bad news or as they think about something painful or negative.

184. CHIN WITHDRAWING—When we are worried or anxious, we instinctively move our chin as close to the neck as possible—nature's way of protecting our vitals. This is an excellent indicator of insecurity, doubt, even fear. If you see this behavior after asking someone a question, there are serious unresolved issues. When children are questioned about something they should not have done, the chin often comes down, showing contriteness. Many adults respond the same way.

185. CHIN HIDING—This is generally employed by children to hide their embarrassment, show their displeasure toward others, or demonstrate that they are upset. They tuck their chin down, often crossing their arms at the same time and then refuse to lift their chin up. In adults, chin hiding is seen between males, standing face-to-face, angry or yelling at each other. In this case it serves to protect the neck in the event of a violent confrontation.

186. CHIN DROP WITH SHOULDERS SLUMPING—This is another behavior familiar to parents—when children lower or try to hide their chin with the shoulders slumped, effectively saying "I don't want to." If the arms are also crossed, then the child definitely does not want to.

187. CHIN TOUCHING—We touch our chins when we are thinking or evaluating something. This is usually done with the tips of the fingers. It is not necessarily a sign of doubt but is something to note when a person is processing information. When coupled with other behaviors, such as lip pursing, it suggests that the person is contemplating something negative, or an alternative to what has been discussed.

188. CHIN BRUSHING WITH BACK OF HAND—In many cultures this signifies that a person has doubts about what is being said. This may also be coupled with lip pursing. It can be performed side to side or from back to front of the chin.

189. CHIN CRADLING—Placing one's chin on the palm of the hand, coupled with relaxation of the facial muscles, suggests boredom. But in a law enforcement context, it might signal a range of possibilities, depending on the circumstances. In a forensic setting, I have seen the guilty strike this pose while sitting in a room alone as a form of perception management, to make authorities think they are so innocent, they are practically bored.

190. ANGRY CHIN PERCHING—This chin perch is performed by placing the chin on the knuckles of the fists, while the elbows are wide and resting on a table as the person stares into the distance or at a computer screen. Usually the forehead is furrowed or the eyes are narrowed or squinting, as a result of something difficult they are pondering or momentary anger. When you see someone posed like this, it is wise to not interrupt.

191. CHIN SHIFTING—Moving the chin left to right against the palm of the hand is a subconscious conveyance of disagreement. I have seen people sitting around a conference room table show their silent displeasure by shifting their chin while resting on the palm of their hand.

192. BEARD/MUSTACHE STROKING—Stroking a mustache or a beard can be highly effective for pacifying stress. As with any repetitive behavior, ignore it if you see it too often, as some people with facial hair do this compulsively. If you see it occur suddenly for the first time or it increases after a topic is mentioned, perhaps the person has an issue. Cultural context must also be taken into account; for instance, beard stroking is common among many men from the Middle East as they pass the time talking. Note that many men with beards find it soothing to stroke their beards as they pass the time of day.

193. **CHIN DIMPLING**—When people are stressed, experiencing emotional turmoil, or about to cry, their chin will dimple. This is true for even the most stoic of individuals.

194. **CHIN-MUSCLE QUIVERING**—The sudden quivering of chin muscles indicates fear, concern, anxiety, or apprehension. People who are about to cry will also do this. The *mentalis* muscle, which covers the chin and causes the skin to quiver, is one of the muscles that most reflects our emotions, according to Dr. David Givens. Sometimes the chin will reflect emotional turmoil even before the eyes.

195. **CHIN TO SHOULDER**—We often see this with people who are embarrassed or emotionally vulnerable. They will, in a very childlike manner, place their chin against one shoulder, looking demure. You should especially note when someone does this while answering a question. It usually means the person has great difficulty discussing a subject, perhaps because she possesses knowledge she does not wish to reveal.

196. CHIN POINTING—In many cultures, people will point in a direction with their chin, extending it forward as they stretch their neck. This replaces pointing with a finger, and is seen throughout the Caribbean, in Latin America, in parts of Spain, and in the Middle East, as well as on many Native American reservations.

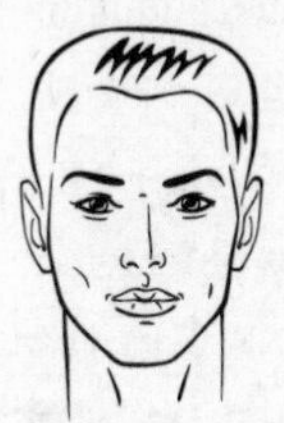

THE FACE

Though I have already covered individual elements of the face, some behaviors are best understood within their full context. Human beings evolved to glean a great deal of information from the face. The eyes and the mouth especially draw our attention. Usually when we look at someone we like, we toggle our gaze between the eyes and mouth, because these two reveal so much information. Mother and baby scan each other over and over to imprint on each other, to collect information, but also to bond—no less so than lovers silently scanning each other in a café. We are naturally fascinated by faces—millions of words have been spent describing the most famous face of all, the *Mona Lisa,* precisely because she is so enigmatic. We are naturally curious about faces, and we are enticed when we see something special in them. Faces communicate emotions, thoughts, and feelings, and so throughout our lives we constantly search there for clues. When the Greeks said that a face "launched a thousand ships," it was both metaphorical and quite likely true—that, too, is the power of the face.

197. **FACE AVOIDANCE**—For a variety of reasons, we sometimes try to avoid face-to-face contact with others, even when we are in their immediate proximity. You see this in court between victim and suspect, or during contentious divorce proceedings. The avoidance becomes obvious by how quickly people will change their demeanor, where they are looking, and how stiff they become, not wishing to look about.

198. **FACE BLOCKING**—This behavior is noted by the person placing their elbows on top of a table and holding their hands together in front of their face. When asked a question, rather than putting their hands down, they peek around their hands or answer directly into them. They are in essence insulating themselves because of stress, lack of confidence, or because they don't like the person they are talking to. The hands serve as a psychological barrier. The reluctance to unveil the face is often a strong indicator that there are issues.

199. **FACE SHIELDING**—Around the world, people will cup their hands over their face or use objects to hide their face, usually as a result of shame, embarrassment, fear, anxiety, or apprehension. Oftentimes when arrestees are being led to the waiting police car, they will use articles of clothing to face shield.

200. EMOTIONAL ASYMMETRY OF THE FACE—Recently it has been shown that the face is remarkable in its ability to reveal multiple sentiments at once. It can sneer and show contempt while at the same time giving a social smile. This is likely evidence of multiple internally competing sentiments, which show up on the face as "leakage." In my observations, the left side of the face (the right side as you look at the person) tends to be more accurate, especially when it comes to negative emotions. This ability of the face to demonstrate different emotions on different halves is called *emotional chirality*.

201. FACIAL INCONGRUENCE—Incongruence between what a person says and how it is reflected in the face is not uncommon. People might say one thing, but their face is already telegraphing another. During an exchange of pleasantries, a very tense face or a face displaying dislike or discomfort betrays true sentiments, though the person might be obliged to say something nice or offer a polite greeting.

202. ODD FACE IN CROWD—In dealing with the United States Secret Service on protection details as well as various private-sector companies over the years, I have learned that in a crowd it is often worth trusting our intuition about the odd face that stands out. By that I mean the one that looks angry when everyone else is happy, or that seems transfixed and rigid when the rest of the crowd is displaying a variety of moods. Airline personnel tell me that in a long queue at the airport, it is the odd emotionally charged face, the one that is not fitting in with the others, that often causes the most problems at the counter.

203. SERENITY IN TURMOIL—Often referred to as "narcissistic serenity," this takes place when the face has an unusual and incongruous expression of calm when the situation would seem to call for anything but calm. Lee Harvey Oswald, Timothy McVeigh, and Bernie Madoff all had this same oddly serene look when arrested, despite their circumstances and the horror of their individual crimes.

204. OUT OF PLACE SMIRK ("DUPING DELIGHT")—This term, coined by famed researcher Paul Ekman, refers to the out-of-place smirk or half smile a person gives when getting away with something. It is very similar to serenity in turmoil (see #203). Duping delight is also seen in those who have outwitted someone, or who think someone has bought in to their lies. It is a pretentious smile at a time and place where humility, seriousness, or even contriteness are more appropriate.

205. FACE TOUCHING—Face touching serves a multitude of purposes. It can attract others—we often see models touching their face on magazine covers. Or it can help us relax by stimulating the myriad of nerves on the face. Context is key.

THE NECK

The neck is the weakest and most vulnerable part of our body. Everything critical for our survival—blood, food, water, electrical signals, hormones, air—flows through the neck. Made up of numerous muscles intricately interwoven to hold up our head, hollow cervical bones that protect the spinal cord, with large veins and arteries that feed the brain, the neck is obviously very vital. And yet, the neck is often ignored when it comes to nonverbal communication, even though we know that our necks signal when we are comfortable, interested, or receptive to an idea or a person. We touch our necks, cover them, or we ventilate them, along with other behaviors and in doing so we tell the world what we are secretly thinking or feeling. Sensitive to the slightest touch or caress or even the warmth of a breath, the neck is also one of the most sensual areas of the body.

206. NECK TOUCHING—Beyond scratching an itch, neck touching serves as an excellent indicator of insecurities, apprehension, anxiety, worries, or issues. However slightly, we tend to touch our neck when something bothers us or we are concerned. Neck touching, in all its forms is often overlooked, and yet, it is one of the most accurate when it comes to revealing that something is bothering us.

207. COVERING OF SUPRASTERNAL NOTCH—Touching or covering of the "neck dimple" or *suprasternal notch* (the indented area of the neck below the Adam's apple and just above the upper chest) indicates concern, issues, worries, insecurities, or fear. Men tend to grab their neck or throat robustly or cover this area with their full hand as they adjust their tie or grab their collar. Women touch this area more frequently than men, and they tend to do so more lightly, with the tips of their fingers. Whether done delicately or strongly, covering the weakest point of the body signifies that something is at issue. Covering our neck when we feel threatened most likely evolved as a result of the countless encounters our species had witnessing countless acts of predation by large felines which usually go for the neck. For more information about this, see *What Every BODY Is Saying.*

208. TIE-KNOT TOUCHING—The tie knot covers the suprasternal notch and the neck, and touching it serves the function of protecting the neck and relieving anxiety. Men will often do this when they feel social awkwardness or mild anxiety. Some men will do this repeatedly as a pacifying behavior, much as a woman might play with a necklace when stressed (see #209).

209. PLAYING WITH NECKLACE—Playing with a necklace serves the same purpose for women as covering the neck dimple with the hand. It protects a vulnerable area and relieves stress through repetitive movement.

210. PLAYING WITH SHIRT COLLAR—Touching or playing with the front shirt collar serves to pacify or relieve stress in three ways: by covering the neck area; as a tactile repetitive behavior; and by moving clothing to ventilate the skin underneath.

211. NECK MASSAGING—People often massage the sides or back of their neck to relieve stress. It is easy for many people to dismiss this sort of behavior, but the fact is that people usually *only* do it when something is bothering them.

212. MASSAGING VAGUS NERVE—The *vagus* (Latin for "wandering") *nerve* connects the brain to our major organs, including the heart. Under stress, you might find yourself massaging the side of the neck, near where you check your pulse. There is a reason for this: stimulation of the vagus nerve results in the release of *acetylcholine,* a neurotransmitter that in turn sends signals to the heart, specifically the *atrioventricular node,* which causes your heart rate to go down.

213. **SKIN PULLING**—Pulling at the fleshy area of the neck under the chin serves to calm some men. Sometimes, under great stress, the pulling becomes extreme. It is rare to see in women. I have seen men under stress pull with such vigor, it makes their skin blanch.

214. **VENTILATING NECK**—When we are under stress, our skin warms, a physiological reaction controlled by our autonomic nervous system and over which we have little control. This often takes place in less than 1⁄250th of a second. By ventilating the collar and neck area we relieve the discomfort caused by the flushing or warming of the skin. Heated arguments or even discussions will cause individuals under stress to ventilate, as will hearing a word or a comment that is hurtful. Those of you familiar with the late comedian Rodney Dangerfield (movie *Caddyshack,* 1980) will remember him doing this in the movie and in his stand-up comedy routines when he didn't "get no respect" but especially when he was stressed.

215. **HOLDING FIST IN FRONT OF NECK**—Placing a fist at the front of the neck serves the same purpose as covering the neck dimple (suprasternal notch). It is an automatic, subconscious response to threats, fears,

or concerns. This behavior occurs primarily in men, but I have seen a few women exhibit it when they are under extreme stress or confronted by something very negative. Many people mistake the fist for a sign of strength, when in reality, in this instance, it is a sign of defensiveness, anxiety, and dislike.

216. **NECK VEINS PULSING**—Noticeable pulsing of veins in the neck indicates stress or anxiety. When a person is fearful or angry, the pulsing can be very noticeable in some.

217. **HARD SWALLOWING**—A hard swallow is highly visible and sometimes audible. It is a spontaneous reaction to something unsavory, dangerous, or extremely stressful, and a reliable indicator of distress. The muscles and ligaments that surround the throat tighten, which causes the Adam's apple to move energetically up and down.

218. **NECK STRETCHING**—Neck stretching or cracking in a circular motion is a stress reliever and pacifier. This is often seen when people are asked difficult questions they would rather not answer.

219. **NECK AND FACIAL FLUSHING/BLUSHING**—Neck and facial flushing is an autonomic response to a stimulus and cannot be controlled. Many people blush

when they feel threatened or insecure and in very rare cases when they are caught lying or doing something illegal. This behavior lets us know that the individual is troubled, whether by merely an innocent embarrassment or something more nefarious. Keep in mind always that certain drugs or foods can cause blushing.

220. **ADAM'S APPLE JUMPING**—If someone's Adam's apple suddenly jolts upward, chances are he's just heard something that has put him on edge, threatens him, or causes apprehension. This uncontrollable reaction also occurs when a person feels highly vulnerable or exposed. The medical term for the Adam's apple is the *laryngeal prominence.* The thyroid cartilage around the larynx (a part of the throat that holds the vocal cords) gives it its protruding shape (prominence). It is usually larger in men than in women. This area of the body is highly sensitive and reactive to emotional stressors.

221. **NECK EXPOSURE**—The canting of the head to the side, exposing the side of the neck, is one of the most used yet least understood body-language behaviors. We instinctively tilt our head when we hold or even see a newborn baby—something the child recognizes and rewards over time with a smile and relaxed face. As we get older, the head tilt features in courtship behavior, as we stare into a lover's eyes with our head canted to the side, exposing our vulnerable neck. In

personal and professional relationships this behavior also signifies that a person is listening and interested. It is a powerfully disarming behavior—extremely useful during a confrontation. Coupled with a smile, this is one of the most effective ways to win others over.

222. **NECK STIFFENING**—When people are attentive and receptive, and especially when they feel comfortable, they will tilt their neck to the side, exposing more of the neck than usual. If the feeling of comfort fades, however, their neck quickly becomes rigid. A stiff neck signifies hyperalertness and vigilance, and might suggest that a person takes issue with something that was just said, or has a serious matter to discuss. When a person goes from a relaxed state to a quick stiffening of the neck, it is a sure sign that something is amiss.

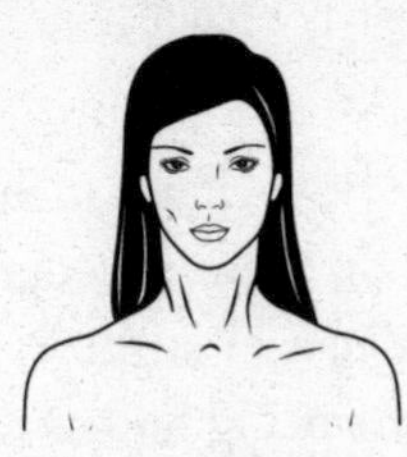

THE SHOULDERS

Whether they are broad, narrow, athletic, slim, attractive, beguiling, or slumping, our shoulders speak volumes about us. Even from a distance, one cannot mistake the broad shoulders of an Olympic swimmer or the sinewy shoulders of a world-class ballerina. The padded shoulders of a business suit make its wearer stand out, just as the bare shoulders of a statuesque model grab our attention. Slumped when we are depressed or wide and pulled back when we're standing proud, shoulders communicate on our behalf. What they say about who we are, what we have achieved, or what we are thinking or feeling will surprise you.

223. RAISING ONE SHOULDER—One shoulder raised toward the ear as a person answers a question usually represents insecurity or doubt. In conjunction with other behaviors (hesitation in answering, arms drawing closer to the body), this is a good indication that the person lacks confidence in what he is saying. In negotiations, when one party raises a single shoulder in response to a query, such as "Is that your best price?," it generally signals there is room for negotiation. A one-shoulder-up answer suggests a lack of full commitment to what is being said.

224. **SHOULDER INTEREST**—The slow, intentional rise of a single shoulder, coupled with a head tilt toward the same shoulder while making direct eye contact, signifies a personal interest. We mostly see this in dating situations, usually in women as they eye someone they like.

225. **HOLDING SHOULDERS HIGH**—When people raise and keep both shoulders high (toward the ears), they are likely feeling insecurities or doubt. This behavior is called the "turtle effect." In essence, the individual is trying to hide in the open. High shoulders are not a sign of confidence. You often see this when a speaker asks for volunteers from a large group or from a student who is ill prepared for a question.

226. **QUICK SHOULDER SHRUGGING**—When people are asked a question and don't know the answer, they often raise both shoulders quickly and prominently. The quick upward movement is a gravity-defying behavior and those are usually associated with positive feelings—in this case they really don't know. This tends to be more honest than a slow shoulder shrug (as they

answer "I don't know") or a shrug of just one hesitating shoulder.

227. **SITTING LOWER AND LOWER**—People who sink lower and lower into their chairs during a meeting are revealing apprehension or lack of confidence. Like the turtle effect, this is a way of hiding in the open—they might also be hoping not to get called on. But note also that in some people this might be simply a sign of indifference or disinterest. This behavior stands out because the shoulders are lower in relation to the table.

228. **SHOULDER/CLAVICLE RUBBING**—During intense or stressful interviews, interviewees will reach across their chest and press their hand against their opposite shoulder, then slowly move their hand across their clavicle toward the chest. Sometimes the hand will then hover pressed against the chest area, or the process may be repeated. The tactile and repetitive nature of the behavior help to relieve stress or apprehension.

229. **SHOULDERS WIDENING**—The amplification of the shoulders from relaxed to broad can be a perceptible display of authority and confidence that signals a person is in charge. We often see this with athletes and military personnel. This is why business suits have padded shoulders—to make their wearers look more powerful and authoritative.

230. **RAISING SHOULDERS WITH PALMS UP, HEAD CANTED**—This is the "Please, why not?" look, performed with the palms of the hands up, head canted to one side, with one or both shoulders up. It is a pleading behavior. Children do this as well as adults, and you often see it with athletes when they want a referee to reconsider an adverse decision.

231. **KOWTOWING**—This is a slight bending forward of the upper torso and shoulders, which may be intentional or subconscious. Around the globe, it is performed in some variant in the presence of a higher authority. In Asia, people bow out of respect, just as the queen's subjects do in London. The origin of kowtowing has much to do with our primate legacy, where everyone bends lower to the alpha male—in our case, someone of higher authority. As a testament to its universality, when the conquistadores reached the New World, they found that Native Americans also bowed or kowtowed to their king, just as they themselves had done in Queen Isabella's court.

THE ARMS

Our arms not only protect us, balance us, and help us carry things, they also communicate extremely well. From our self-hugs when we are stressed, to the raised arms of a person who just came in first place, to the outward reach of a child seeking a loving hug, our arms are continually assisting us, warming us, attending to others for us, and communicating our needs as well as how we feel—far more than we realize.

232. **HUGGING**—Hugging, in all its forms, is universally indicative of closeness, good feelings, warmth, and co-operation. While in some cultures a brief social hug, an *abrazo* (in Latin America), can serve as a greeting gesture similar to shaking hands, how it is performed can indicate how the participants feel about each other. Consider American athletes and movie stars giving each other bro hugs. As an observer, I always note the hug and the facial expressions to give me a reliable sense of how two individuals really feel about each other.

233. ANIMATED GESTURES—Animated gestures reflect our emotions and also get us noticed. Broad gestures are powerful displays when we're speaking and are essential to dynamic communication. In many cultures, emphasis requires exaggerated gestures. To an outsider, people making such gestures might look like they're about to fight when in fact they are just being emphatic.

234. GESTURING WHILE SPEAKING—I often get the question "Why do we gesticulate?" Gestures are considered an integral part of communication. Gestures help us to get and maintain attention as well as to highlight important points. Gestures even help the person speaking by facilitating greater flexibility in speaking and even with the recall of words. Gestures affect how our message is received and how much of it other people remember. When gestures echo the message, the message is potentiated. We want to be seen gesturing as we speak. If you look at successful TED Talks, you will notice that gestures are an essential element utilized by the best speakers.

235. ARMS AGAINST BODY, HANDS FLEXED—This is often referred to as restrained elation. When people are pleased with themselves but are trying not to show it, they might hold their arms against their body and then lift their hands at the wrist so the wrist is almost at a ninety-degree angle, with the palms facing down. This can also take place when people are trying to control their excitement and don't want to be noticed. The behavior may be accompanied by a rise in the shoulders and of course facial displays of joy.

236. **ELATION/TRIUMPH DISPLAYS**—Displays of elation or triumph tend to be gravity-defying—in other words, the gesture is made upward or outward away from the body. Sometimes we actually jump out of our seats into the air with our arms and fingers extended. Positive emotions drive gravity-defying gestures, and so around the world triumph displays at sporting events tend to be similar: arms up in the air.

237. **ARMS HELD BEHIND BACK**—The regal stance is performed by placing the arms and hands behind the back. Queen Elizabeth, Prince Charles, and other British royals often walk this way when they want others to remain at a distance. For the rest of us, too, it signals to others to give us space. It is not a good way to endear yourself to others as we tend to associate aloofness with this behavior. Interestingly, young children don't like when their parents hide their hands behind their back.

238. **ARMS STIFFENING**—People's arms will frequently stiffen when they are scared or overwhelmed by an

event. Their arms lie dormant at their sides, making them look unnatural or robotic. Stiff arms are a strong indicator that something negative has just transpired.

239. **ARMPIT EXPOSING**—The exposure of our inner arm, including the armpit (*axilla*), is reserved for those times we are comfortable around others. Women especially might use this behavior (scratching the back of the head while exposing the axilla directly toward a person of interest) in order to garner that person's attention and demonstrate her interest. Conversely, when our armpits are exposed and someone comes near that makes us feel uncomfortable, we will immediately cover our armpits.

240. **ARM CROSSING/SELF-HUGGING**—Self-hugging is an effective way to comfort ourselves while waiting for someone to arrive, while watching a movie in public, or when we need a little bit of self-comforting. This explains why so many passengers on a plane will cross their arms while standing in line to use the restroom. We cross our arms for many reasons. Here are some of the reasons reported to me: "It's comfortable"; "It's useful when my arms are tired"; "It hides my boobs"; "I do it when I am inquisitive"; "It hides by belly." Everyone has a good reason and most of the time, it gives them comfort. There are a lot of people that mis-

takenly equate crossing the arms with keeping people further away—that is usually not the case.

241. **ARM CROSSING/PROTECTION**—In some instances the arm cross is a means of protection, rather than a comforting gesture. We might subconsciously seek to shield our vulnerable ventral (belly) side when we feel insecure or threatened. In those cases, we will see more tension in the arms and psychological discomfort in the face.

242. **ARM CROSSING/SELF-RESTRAINT**—People might cross their arms to restrain themselves when they're upset. Picture a customer at an airport counter who has been bumped from a flight. Whereas the self-hug (see #240) is done with very little pressure, this behavior helps to literally restrain the arms as emotions get out of control. Note that this self-restraining behavior is usually accompanied by facial displays of animosity.

243. **ARM CROSSING/DISLIKE**—In the presence of someone we don't like, we might draw our arms across our belly, attempting to distance or insulate ourselves from that person. Usually this occurs as soon as we see someone objectionable, and that is what distinguishes this behavior and communicates our dislike

very accurately. This should be differentiated from self-hugging behaviors by other cues that accompany it, such as a tense face and feet that also turn away.

244. **ARM CROSSING/MASSAGING**—Crossing the arms at the chest can be comfortable for many people. However, massaging the opposite shoulder or arm suggests that a person is stressed or concerned. This is most likely to occur when the person is seated at a table with her elbows on the surface, but I have also seen it in people sitting in a chair, a form of self-hugging while they massage the opposite arm to relieve stress or worry.

245. **ARM CROSSING, HOLDING WRIST**—When confronted with damaging information in a forensic setting, interviewees will suddenly reach across their belly and hold the wrist of the opposite hand while sitting. Look for it immediately after a person has been asked a difficult question or is accused of something. Poker players have been observed displaying this behavior when their hand is weak or marginal.

246. **ARM SPREADING**—People who spread out their arms over several chairs or a couch are demonstrating confidence through a territorial display. Senior executives will do this more often than junior staffers. Observe when someone of higher rank or status walks in whether the person withdraws his arms to his sides.

247. **ELBOWS SPREADING OUT**—When people are strong and confident, they will gradually take up more space, spreading their elbows across a table or desk. This tends to be subconscious, and they are generally not aware that they are publicizing their self-assuredness.

248. **ELBOWS NARROWING**—When we're sitting with our arms on a table, the moment we feel insecure or threatened, we will narrow our elbows on the table. We can use this metric to help us assess how committed or confident others are as different topics are discussed.

249. **ELBOW FLEXING**—The elbow flex is performed by placing hands on the hips, arms akimbo, and flexing the elbows forward (like a butterfly flapping its wings) each time we want to emphasize what we're saying. This is a territorial display that also projects confidence. I have seen senior managers, coaches, and military officers do the elbow flex as they emphasize a particular point.

250. **ELBOWS INTERLOCKING**—In many parts of the world the interlocking of arms at the elbows with another person as you walk or sit is a sign that you are close to the person or that you are having a very private conversation. This behavior draws the hips close together, which suggests that things are going well. It is not unusual in Mediterranean countries or

in South America to see both men and women walking arm in arm.

251. WRIST BEHAVIORS—We might not think of the wrists as a window into the mind, but they can be. We expose the underside of our wrists to others when we like them or feel comfortable around them. Holding a drink or a cigarette, a woman will expose the inner wrist to a nearby person if she is interested in them or comfortable. The minute she is not, she will rotate the wrist and only expose the outside of the wrist. Our limbic system protects us by orienting our most sensitive areas—the underside of our arms, our neck, our bellies—away from those we dislike or find threatening.

252. GOOSE BUMPS—Also called "goose pimples" or "gooseflesh," this is an involuntary reaction to cold or perhaps even fear—usually visible on the arms and legs. The formation of goose bumps causes hair to stand up on the surface of the skin, which is why the medical term for it is *horripilation* or *piloerection* (see #253). In primates, this display is even more noticeable when they are scared, as their hair stands up to make them automatically look larger. Because we as a species have lost most of our hair, we only see the remnants of piloerection through goose bumps.

253. **HAIR ERECTION (PILOERECTION)**—Sometimes the hair on the arms, torso, or back of the neck will stand up visibly. From an evolutionary perspective, this is believed to be a vestigial response we share with primates to make us look bigger when we are scared, frightened, or fearful. When we subconsciously assess a person, a place, or a situation as potentially dangerous, the hair on the back of our neck will stand up—when you feel this, take note. These subconscious sentiments of ill feeling or danger, according to Gavin de Becker in his book *The Gift of Fear*, should not be ignored.

254. **EXCESSIVE SWEATING**—People under stress may suddenly sweat profusely as their body attempts to ventilate itself through evaporation. Many a drug trafficker has been stopped at the border because he is the only one with sweat rings around his armpits and his neck glistens with moisture when he pulls up to the customs officer. Excessive perspiration may signal that a person is hiding something or is about to commit a crime. That doesn't mean every sweaty person is guilty of something—just that it behooves us to pay closer attention.

255. **SELF-INJURY**—Individuals who suffer from borderline personality disorder, as well as others who are emotionally unstable or depressed, might bear scars where they

have cut, slashed, or burned themselves intentionally. Recognizing these signs in others is key to getting them help. They might not seek help themselves, but they are nonverbally communicating their mental health needs through self-injury.

256. NEEDLE TRACKS—Individuals who use heroin and other intravenous drugs will have scars tracking their veins on the inside of their arms. On long-term abusers, this can be very evident.

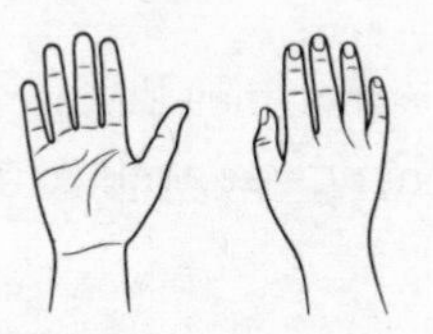

THE HANDS AND FINGERS

The human hand has no equal. It can hold a scalpel and perform delicate surgery or grasp a brush and paint the ceiling of the Sistine Chapel. Hands gently cradle us at birth and just as easily grasp an axe handle with enough force to chop down a tree. Our hands are indispensable for work, for play, and for our protection, and we rely on them every day to interact with the world around us. We also use them to communicate effectively—whether we are stopping traffic at a school crossing, conducting an orchestra, or signaling a friend to quickly come over. Our hands are constantly communicating to others our passions, our desires, our abilities, our concerns, and most important, through the gentlest of touch, our love.

257. **CONDITION OF HANDS**—You can tell a lot from a person's hands. Grooming, scars, and calluses might indicate what kind of work the person does; an office worker's hands look quite different from a cement mixer's. Similarly, arthritis as well as neurological

disorders can sometimes be discerned from the condition of the hands as well as finger movement or agitation.

258. **HAND GROOMING**—Well-groomed hands are a sign of a healthy individual. Clean fingers, with nails of proper length, indicate that people care about themselves. This contrasts with people who have dirty or long nails, unkempt cuticles, or flesh that has been chewed off the fingers. In dating and even in the workplace, we often draw conclusions from how well-groomed or healthy-looking a person's hands appear.

259. **FREQUENCY OF TOUCH**—How frequently we touch others is a good way of communicating how we feel about them. While on some level the degree of touch is culturally determined, for the most part when we care about others we tend to touch them more often.

260. **HOW WE TOUCH**—A touch can be respectful, loving, playful, sensuous, reserved, tender, caring, or palliative. A light touch of the skin can send shivers down our spine, stimulating sexual desire. In fact, a light touch stimulates the brain differently than a heavier touch. The loving touch of a caring person with a full palm, warm from the presence of blood near the surface of the skin, can communicate so much to a newborn as well as a lover. And yet when our boss pats us

on the shoulder with just the fingertips and says "Good job," our skin crawls because the gesture doesn't feel right; we know it is contrived and empty.

261. **TOUCH AND SOCIAL STATUS**—In most cultures, whom we can touch, and how, is dictated by social convention. In almost all societies, higher-status individuals will more often touch lower-status individuals than the other way around. In a work environment, you are more apt to see the boss giving a pat on the shoulder than to see an employee pat the boss. We must also be aware of when it is proper to touch, where it is proper to touch (arm or elbow, for instance), and when or if it will be appreciated.

262. **PRESSING THE FLESH**—This is a term that is often used by politicians to mean shaking hands, gripping an arm, hugging, holding or kissing babies. Hand shaking can be an opportunity for politicians to humanize themselves and establish physical bonds. The connection is literally chemical, as *oxytocin* (a powerful hormone that serves to socially bind us) is released when we touch others.

263. **HANDS ON HIPS, ARMS AKIMBO WITH THUMBS BACK**—Known as arms akimbo, when the hands are on hips, thumbs back, with elbows out, this is a dominance display. This signals that an individual is ready

and alert, has something to discuss, or there is an issue. Airline agents tell me that if a person does this while waiting in line, you can bet he has a complaint. This is a very authoritarian look. This is not a good way to talk to one's children—it inhibits communication, as the parent looks like a military drill instructor.

264. **HANDS ON HIPS, ARMS AKIMBO WITH THUMBS FORWARD**—This is more of a curiosity display. The position of the thumbs may seem a minor detail, but it is significant. Witnesses often stand and contemplate an event this way, while those who take action (police, firefighters) will do so with the thumbs back.

265. **TERRITORIAL HAND DISPLAY**—As a form of mild intimidation, a person will splay out their hands on a desk or table. You see this often at a return counter, where angry customers take up increasingly more space with their hands as they argue with the representative. As emotions increase, note how the hands grow further and further apart.

266. **PUSHING AWAY FROM TABLE**—A sudden stiff-arm pushing away from the table is a very accurate indicator the person disagrees with something said or discussed or might feel threatened. The speed of the motion matters: the more quickly it happens, the more concerning.

267. PLAYING WITH OBJECTS—Playing with jewelry or other objects (winding a watch, tapping a pencil, checking a smartphone) serves as a pacifier. You often see this behavior in people waiting to be interviewed for a job or to just pass the time. This is different from "surrogate touching" (see #291).

268. OBJECT PLACEMENT—We might surround ourselves with objects—whether a pencil and paper on a work desk or a jacket on a theater chair—to establish our territory. Object placement can also signal that we are not fully interested in someone or that a relationship is in trouble. At a restaurant for instance, when things are good, we tend to move objects out of the way to get a clearer view of our companion; when they're not, we'll place flowers or a drink bottle in our line of sight to serve as a barrier across the table. It is especially telling when someone moves objects while speaking.

269. HAND STEEPLING—The hand steeple is performed by placing the fingertips of both hands together, spreading them, and then arching the hands so that the tips of the fingers look like a church steeple. This is a universal display of confidence and is often used by those in a leadership position. Chancellor Angela Merkel of Germany is famous for steepling often. Note, however, that confidence doesn't always guarantee accuracy. A person may be wrong in their facts but confident as they speak. Nevertheless, steepling is a useful gesture to convince others of your commitment to what you are thinking or saying.

270. **MODIFIED STEEPLING**—The modified steeple is performed by interlacing all of the fingers, with the exception of the index fingers, which are erect and touching at the fingertips. It looks more contrite than a regular hand steeple; nonetheless, it still signifies assurance and confidence.

271. **HANDS IN READY/ACTIVE POSITION**—This is where the hands are held about fourteen inches apart in front of the belly at waist level with the palms facing each other, fingers spread apart. Speakers often do this to capture an audience's attention at an important moment. This is not the rogatory position (see #272), which requires the palms to be facing up; here the palms face each other as if the person is holding a beach ball. This is a useful gesture to build into your public-speaking repertoire.

272. **PALMS-UP DISPLAY**—Also known as the *rogatory hand position,* this is a universal behavior of humility, compliance, or cooperation, used by people who want to be accepted or believed. The presentation of the palms of the hands facing up is a universal way of saying "My hands are clean," "There is nothing hidden

here," "I implore you," or "I am at your command." It is also used in religious ceremonies to demonstrate humility and piety.

273. **PALMS-DOWN DISPLAY**—Palms-down displays are more affirmative than palms-up displays. These might be made on a table or symbolically in the air. The farther apart the arms are (in a two-handed gesture), or the harder the hand slaps down, the more committed the person. Affirmative declarations such as "I didn't do it," when spoken as the palms push strongly downward on a table, tend to have greater validity. Liars struggle to do this properly, generally performing the gesture too passively.

274. **PALM DOWN, FINGERS SPREAD**—When a person makes a formal declarative statement, such as "I didn't do it," with palms firmly placed down and fingers spread wide, it is more likely to be an authentic answer. I have never seen a liar successfully pull off this gesture, probably because the thinking part of the brain is out of sync with the emotional part of the brain. In other words, they know what to say—"I didn't do it"—they just don't know how to dramatize it because the emotional side of the brain is not fully committed.

275. **HAND RESTRICTING**—Researchers, Aldert Vrij in particular, have noted that when people lie, they tend

to use their hands and arms less. This can be a powerful behavioral marker, though it can simply indicate shyness or discomfort. This is where having a baseline of the person's normal behavior is so important. In any case, it is a behavior to note but not to necessarily equate with deception.

276. **HAND WRINGING**—Rubbing one's hands together communicates concern, doubt, anxiety, or insecurity. The degree of stress is reflected in how tightly the hands are wrung. Blotches of red and white skin on the fingers or hands indicate an elevated level of discomfort.

277. **FINGER HOLDING**—When we meet people for the first time or we feel a little insecure, we tend to hold our own fingers together lightly in front of us. It is a very tactile, self-soothing behavior. Prince Harry is famous for this but we all do it as we patiently wait in line or speak to someone we've never met before.

278. **JITTERY HANDS**—When we are excited or stressed our hands may become jittery. Jitters, of course, can also be caused by a neurological disorder, disease, or drugs, but for the most part, when a person appears otherwise healthy, we should take notice. People might accidentally knock down objects such as wineglasses when stressed, or their spoons will

shake in their hands. Fingers and hands might shake uncontrollably after an accident or when we are notified of terrible news.

279. **HANDS AS ANCHORS**—This is where we take possession of an object to let others know it is ours. It might also happen with other people, as when we're talking to someone we like and use our hand as an anchor near this person so others will stay away. You see this most often at bars or parties—men will pivot around the anchor point as if permanently attached in order to make sure others don't intrude. It is a territorial display.

280. **HAND THRUST TO FACE**—This might come as the final affront in an argument. A raised palm thrust at the other person's face says to stop, go no further, or, in the common parlance, "talk to the hand." This can be a very insulting gesture and certainly has no place in amiable interpersonal communication and certainly not in business.

281. **SELF-TOUCHING WHILE ANSWERING**—Take notice of people who while answering a question are pacifying (any hand-to-body touching or stroking) rather than emphasizing with their gestures. Over the years I've noted that these individuals are less confident than those who while answering use their hands to illustrate a point.

282. **INTERLACED FINGERS—THUMBS UP**—Statements made with thumbs up while the fingers are interlaced indicate confidence. Usually people do this with their hands on their lap or on top of a desk or table; their thumbs rising as they genuinely emphasize a point. This is a very fluid behavior that might change depending upon the emotions felt in the moment, as well as how committed the person is to what she is saying.

283. **INTERLACED FINGERS—THUMBS DOWN**—Fingers interlaced with thumbs down tend to show a low degree of confidence or negative emotions about what is being discussed. When we are really confident about what we are saying, we tend to elevate the thumbs subconsciously. As stated above, this is very fluid, a person's thumbs may go from up to down during a conversation depending on how they genuinely feel about a topic.

284. **THUMB MASSAGING**—Thumb massaging is a mild pacifier. The hands are intertwined and the thumb on top rubs the one below it repeatedly. We usually see this when people are waiting for something to happen, though they might also do this as they are talking, if they are slightly nervous or anxious.

285. **THUMB TWIRLING**—Twiddling our thumbs is a way to pass the time or deal with small amounts of stress. The repetitive nature of it is soothing to the brain.

286. **FINGERS CLOSE TOGETHER**—When we feel concerned, bewildered, humbled, scared, or cornered, we subconsciously make the space between our fingers smaller. In the extreme, when we are very concerned, we curl up our fingers so they are not sticking out. Here, our limbic brain ensures that our fingers are not loose when there is a threat.

287. **THUMB OUT**—When we feel confident, the thumb will move away from the index finger. This is easily observed when hands are on a table. In fact, distance of the thumb from the index finger can serve as a gauge to a person's confidence level. It might also show a person's level of commitment to what she is saying: the greater the distance, the stronger the commitment.

288. **THUMB WITHDRAWING**—When we feel insecure or threatened, we will withdraw our thumbs subconsciously and tuck them next to or underneath the fingers. Doing this suddenly means the person is concerned, worried, or threatened. This is a survival tactic, similar to dogs tucking their ears down to streamline themselves in case of the need to escape or fight.

289. **THUMB DISPLAYS IN GENERAL**—Watch for individuals who display their thumbs as they hold on to a jacket lapel or pant suspenders. I see this often in court performed by attorneys. As with other thumbs

up displays it typically means the person is confident in what they are doing, thinking, or saying.

290. THUMB UP OK SIGN—This, of course, is a very positive sign in the United States, signaling that all is fine. At one time it was used routinely to hitch a car ride. Note that in some cultures, such as the Middle East, a raised thumb is a phallic symbol and should be avoided.

291. SURROGATE TOUCHING—Sometimes, early on in a romantic relationship, we want to be in closer physical contact with the other person but feel it is too soon. So we transfer those wishes to an object. We might stroke our own arm or slide our hand around a glass repeatedly. Surrogate touching is a form or subconscious flirting as well as a stress reliever that often serves as an effective substitute for the touch we desire.

292. RECIPROCAL TOUCHING—This is where someone reaches out to touch us and we touch back in return. Usually it is a sign of social harmony and comfort with others, so when it isn't reciprocated, there may be an issue. Often in work relations, when someone is about to get demoted or fired, there will be less reciprocal touching on the part of the supervisor in the days prior to the employee dismissal. This also happens in dating situations when there is about to be a breakup.

293. **HOLDING ON TO FURNITURE**—If people hold on to their chair, or the edge of a desk or podium, as they make a declarative statement, they are communicating doubt and insecurity. I have sometimes seen this when people sign a contract they are reluctant to endorse but must. As an observer, you should always question what insecurity is driving this behavior.

294. **CLINGING BEHAVIORS**—When children are under stress, they will grab the clothing of the nearest relative for comfort. In the absence of a parent or an object, they will also grab their own clothing as if it were a security blanket—which in essence it is. This tactile experience can be very psychologically soothing. Adults sometimes do this, too, perhaps as they get ready for a job interview or a speech. The great tenor Luciano Pavarotti held a handkerchief in his hand while performing, which, he said in interviews, gave him "security" and "comfort."

295. **EMPHASIZING WITH HANDS**—When we are comfortable, our hands naturally gesture and emphasize. In some cultures, especially around the Mediterranean, people tend to gesture more emphatically, and

these gestures are highly significant in context. Great speakers also gesture frequently. Researchers tell us that when people suddenly begin to lie, they engage in fewer hand gestures—and with less emphasis. If the hands suddenly become passive or restrained, it is likely that the person is losing confidence in what he is saying, for whatever reason.

296. GIVING THE FINGER—Pioneering psychologist Paul Ekman first noted how individuals who harbor animosity toward others will subconsciously give "the finger" (the indecent finger is usually the middle or longest finger as in "F—— you!") by scratching their face or body with it, or even just pushing their eyeglasses back into place. It is a subconscious sign of disrespect.

297. FINGER POINTING—Almost universally, people dislike having a finger pointed at them. If you have to point, especially in a professional or romantic setting, use the full hand, fingers together, rather than a single finger. This also applies when pointing to objects. When directing someone to a chair, do so with the full hand rather than with a single finger.

298. FINGER JABBING—Jabbing a finger at someone's chest or face is a highly antagonistic behavior, used to single a person out when there are issues. When actual physical touch is involved, it becomes even more threatening.

299. USING FINGER AS BATON—This is where the index finger is used to keep a rhythm in speech, cadence, or music. It provides emphasis when it follows a point in speech. It is seen more often in Mediterranean countries, and some people take offense at that "wagging" finger because they don't understand that it's a cultural trait, used for emphasis, not necessarily an antagonistic behavior.

300. TWO-HANDED PUSH BACK—We usually see this when people are speaking publicly. They will hold both hands up in front of them, palms toward the audience, and figuratively push the audience away. This has a subconscious negative connotation as when someone says "I know how you feel," while in essence gesturing "Go away."

301. NAIL BITING—Nail biting or cuticle biting is a way of relieving tension and anxiety. It is a display of worry, lack of confidence, or insecurities. Even people who never bite their nails might suddenly find themselves doing so when undergoing extreme stress. This behavior can become pathological to the point of damaging the skin or even ulcerating the fingers, destroying the surrounding cuticle or otherwise healthy tissue.

302. FINGER STRUMMING—Strumming one's fingers on a table or a leg passes the time and, like other repetitive behaviors, soothes. In professional settings you see this as people wait for someone to show up or finish talking. It is a way of saying, "Come on, let's get things moving here." This is similar to cheek strumming (see #170).

303. HANDS IN POCKET—Many people are comforted by placing one or both hands in their pockets while talking to others. But sometimes this is seen as too informal and in some cultures is considered rude. It should be noted that some people erroneously view keeping hands in pockets as suspicious or deceptive.

304. MASSAGING CLOSED FIST—Massaging the closed fist with the other hand is a self-restraining and pacifying behavior. It usually means the person is struggling or worried and experiencing a lot of underlying tension. You often see this with poker players and stock traders, or wherever fortunes might be quickly won or lost.

305. SPEAKER'S FIST—Sometimes we will see a speaker make a fist while "hammering home a point." This is not unusual, especially from very dramatic or enthusiastic speakers. What is unusual is watching as someone waiting his turn to speak turns his hand into a fist. This usually indicates pent-up issues, constrained energy, or anticipation of some sort of physical response. It is said that Theodore Roosevelt, a dynamo of action and adventure, always sat with his hands balled into fists, as if holding back coiled energy.

306. **RUBBING HANDS ON PALMS**—Rubbing our fingers across the palm of the hand is a pacifier. When it is done repetitively, or with increased pressure, there is high anxiety and concern. We can rub the palm either with the fingertips of the same hand or against the opposite hand.

307. **TEEPEE FINGER RUB**—When people feel concern, stress, anxiety, or fear, they might pacify themselves by rubbing their straightened interlaced fingers back and forth against one another. The interlaced fingers provide a greater surface area to stimulate as the hands and fingers are moved back and forth relieving tension. This is one of the best indicators that something is very wrong or someone is severely stressed. This behavior is usually *reserved* for when things are especially bad. In less dire situations, we will instead wring our hands or rub them together without interlacing the fingers. What makes this behavior stand out is that the fingers are ramrod straight and interlaced.

308. **INTERLACED FINGERS, PALMS UP OR PALMS DOWN**—This is an extreme variant on interlacing

the fingers to displace stress. Here the person holds the hands palms up and interlaces the fingers, pulling the hands upward toward the face and making an awkward-looking triangle, with the elbows down and the palms of the hand arching upward. Or, in the palm-down variant, the palms remain face down and the fingers are interlaced in front of the crotch as if to crack the knuckles. This contortion of the arms and the fingers, by stressing muscles, joints, and tendons of the hand, relieves stress. I have seen this after a teenager crashed his parents' car as he waited for his mother to come pick him up.

309. KNUCKLE CRACKING—Knuckle cracking, in all its varieties, is a pacifying behavior. The act of knuckle cracking for some people seems to soothe tension and so we see it when they are tense or nervous or even bored. People might crack each knuckle individually or all the fingers of one hand at once. This behavior increases in frequency with stress.

310. KNUCKLE CRACKING WITH INTERLACED FINGERS—This behavior is performed by intertwining the fingers, with the thumbs down, and then stretching the arms forward until the knuckles crack. As with similar contorted displays, it signifies a high degree of psychological discomfort, stress, or anxiety. It also serves as a double pacifier: both interlacing the fingers and cracking the knuckles. This behavior is generally exhibited more often by men.

311. **TAPPING SIDES OF LEGS**—People will tap their legs with the palms of their hands (usually near the pockets) when they are impatient or becoming aggravated. I see this all the time in people waiting to check in at hotels. The tactile nature and the repetition make this act both a distraction and a useful pacifier.

312. **PREENING**—It is not only birds that preen. Preening can take many forms: adjusting a tie, repositioning a bracelet, smoothing out a wrinkle on a shirt, fixing one's hair, reapplying lipstick, plucking an eyebrow. We preen when we care enough to want to look our best. Hair preening when we are interested in someone romantically is especially common. The repeated stroking of the hair also gets us noticed. Interestingly, when attorneys do something so simple as pulling at their jacket (a preening behavior) as the jury enters the room, they are subconsciously perceived by jurors as more likable.

313. **PREENING (DISMISSIVE)**—There is another kind of preening, intended to be dismissive or disrespectful—almost the opposite of what I just described. The act of picking lint or hair from clothes or cleaning one's nails when being addressed by another person is inconsiderate at best, disrespectful, even contemptuous, at worst.

314. HAND ON LEG, ELBOW OUT—Sitting with a hand on the leg, elbow out, usually indicates high confidence. As this behavior comes and goes as people converse, we can observe a person's self-assurance waxing and waning. The elbow-out posturing is a territorial display.

315. FINGERS CURLING, NAILS FLICKING—Often when people are nervous, agitated, or stressed, they will curl up their fingers (usually on one hand) and flick their nails against the thumb. They might flick one finger or use a variety of them. It is a way to pacify oneself and can be both distracting and noisy for others.

316. HAND SHAKING—The handshake is *the* favored greeting behavior in the West, appropriate in both professional and personal settings. A handshake is often the first physical contact and impression you will make and take away from another person, and so it is important to get it right. Think of how many times you have received a "bad" handshake (too strong, too wet, too soft, too long). A bad handshake leaves a negative impression that can last in our minds for a long time and make us reluctant to shake hands with that person again. We should remember that the custom of shaking hands is not universal; in some cultures a bow or a kiss on the cheek might be more appropriate. Nevertheless, a good handshake begins with good eye contact, a smile if appropriate, and the arm extended with a slight bend at the elbow. The fingers approach the other person's hand pointing downward, the hands clasp with equal pressure (no one is impressed that you can crush walnuts bare-handed), engulfing each other—this allows for the release of the hormone oxytocin (furthers social bonding)—and after a second or so the hands are released. Older people will require less pressure, and higher-status individuals will set the tone for how long you will shake hands and how much pressure to apply.

317. TENDERED HANDSHAKE—In some cultures, most notably in parts of Africa, it is customary to greet a revered or important person by holding the outstretched right hand supported underneath the forearm by the left hand. The hand is literally being tendered or offered as if it were something precious, in the hope that the person to whom it is offered will take it, thus honoring the offeror. This gesture might look odd at first to Western observers, but it is a gesture of deference and high respect and should be accepted as such.

318. NAMASTE—In this traditional Indian greeting, the hands are placed palms together directly in front of the chest, fingers pointed upward, elbows out, sometimes followed with a small bow or forward lean and a smile. It is used as a formal greeting—in a sense it replaces the handshake—and can also be used to say “so long.” This gesture has a deeper meaning than the Western handshake and must be received with respect.

319. **HAND HOLDING**—Hand holding is an innate human tendency; we observe children doing it very early on, first with parents and later with playmates. In romantic relationships, both its frequency and its type (whether a handclasp or the more intimate and stimulating interlaced fingers) might signal how close or serious a partnership is. And in some parts of the world, including Egypt, Saudi Arabia, and Vietnam, it is very common to see men holding hands as they walk together.

320. **OK SIGN (PRECISION CUE)**—When talking about something very precise, speakers will hold the tip of the index finger and thumb together to make a circle—what we in America call the OK sign. This gesture is very common throughout the Mediterranean and is used to emphasize a specific point while speaking. In the United States we also use this gesture to indicate agreement or that things are fine or OK. Note that in other countries, such as Brazil, this sign can be erroneously interpreted as a vulgar display connoting an orifice.

321. **POLITICIAN’S THUMB**—When politicians are speaking, they will often extend their arm toward the audi-

ence or up in the air while pressing the thumb against the curled index finger to make a precise, strong point. In essence, this is a modified precision grip. Again, we see this more in the United States than in other countries and so it is in part cultural. Bill and Hillary Clinton, Barack Obama, and Canadian prime minister Justin Trudeau are all known for this gesture, usually used when making or emphasizing a specific point.

322. **RING PLAYING**—Playing with a wedding band by twirling it or taking it off and putting it on—is a repetitive behavior that people sometimes used to calm their nerves or pass the time. It is not, as some people claim, an indication of marital unhappiness. It is merely a self-soothing repetitive behavior.

323. **DISTANCING FROM OBJECTS**—When we have negative feelings toward something or someone, we often subconsciously attempt to distance ourselves. People on a diet may push a bread basket a few inches further away at dinner or even ask that empty wineglasses be removed from a table if they dislike alcoholic beverages. I have seen criminals refuse to touch a surveillance photograph or push it back across the table because they recognize themselves in the image. These are important behaviors to note because they speak to what is uppermost on that person's mind at the moment.

324. RELUCTANCE TO TOUCH WITH PALM—The consistent reluctance of a parent to touch a child with the palm of the hand can be a sign of significant issues—whether indifference toward the child or some other form of abnormal psychological distancing. And when couples stop touching each other with the palms of the hand, instead relying on their fingertips, it's likely that there are issues in the relationship (see #260).

325. ERRATIC ARM AND HAND MOTIONS—Sometimes we are confronted by an individual making erratic motions with the arms and hands. The arms and hands might be out of synchrony with the rest of the body and with the person's surroundings. In these instances, the best we can do is recognize that there may be a mental condition or disorder at play. Recognition and understanding are key to lending assistance if necessary, and not to look on as if at a spectacle.

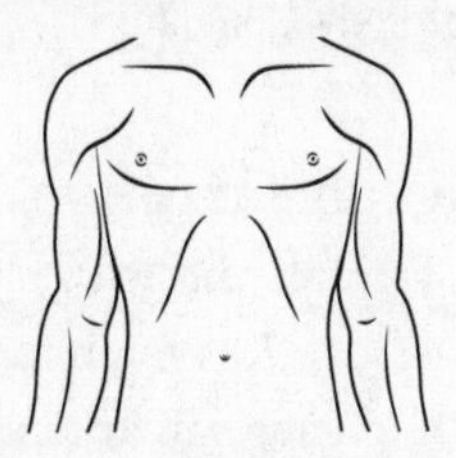

THE CHEST, TORSO, AND BELLY

The torso is home to the majority of our vital organs, is generally our largest body part by mass, and is the area of the body we tend to cover up first when we feel threatened. It is the body's billboard, offering clues (with the help of our clothing) about who we are, what group we belong to, what we do for a living, even how physically fit we are. And of course, much of the body's essential organs—the heart and lungs among them—are located in the torso. Our torso, while rarely recognized in the study of nonverbals, is actually a very good place to collect information about people, from life choices to how they feel.

326. **CHEST HEAVING, RAPID BREATHING**—A heaving chest and rapid breathing usually indicate stress, concern, fear, worry, or anger. Context of course is important as there are many reasons for this behavior, including age, recent physical exertion, anxiety, or even a heart attack. What is important is to observe and be ready to take action if necessary.

327. SHALLOW, FAST BREATHING—Breathing that is shallow and fast usually indicates fear or anxiety, perhaps even a panic attack. Watch for the shallowness of a person's breath to gauge their anxiety level. The shallower and more rapid the breathing, the greater the distress. Useful is to get them to take one long inhale then exhale for as long as possible (3 to 5 seconds) then repeat. This helps to lower the breathing rate.

328. PRESSING ON CHEST—In tense situations, a person will press on their own chest/diaphragm area with the thumb and middle finger (sometimes all of the fingers) in order to relieve sudden pent-up stress. The self-administered pressure on the *solar* or *celiac plexus,* near the center of the chest, which is rich in nerves, seems to have a pacifying effect when pressed upon. The pressure can be very light or extremely forceful depending on the needs of the individual. It is not unusual for someone receiving horrible news to press on their own chest.

329. CLAVICLE MASSAGING—Under stress, individuals will massage the clavicle (collarbone) on the opposite side (e.g., placing their right hand upon their left clavicle). The arm drawn across the center of the body provides a feeling of protection, while the repetitive touching of the clavicle acts to soothe the person. This area of the body is very sensitive to touch—one reason it is considered an erogenous zone.

330. **REPETITIVE HAND RAKING OF CHEST**—Repetitive massaging of the upper chest area with fingers and thumb moving rake-like back and forth is usually a good indicator of insecurity, concern, or issues. This behavior is an extremely reliable indicator of anxiety or even a pending panic attack. What stands out with this behavior is how the curled-up fingers are used, like a claw or a rake, to perform this behavior rather than using the full palm of the hand.

331. **PALM ON CHEST**—In many cultures, people place the palm of the hand on the chest to convey sincerity and as a gesture of goodwill when meeting others. In my experience, both the honest and the deceptive will do this behavior and so we should treat the behavior as neutral. It is neither evidence of honesty nor of sincerity though it may be offered as such. In a forensic setting, if someone says "I didn't do it," as they place their palm on their chest, it should not receive any greater weight or value no matter how well performed. Having said that, I have noted over the years that truthful people tend to press with more force, fingers wider apart, and with the full palm against their chest, whereas those who are attempting to deceive tend to make contact primarily with their fingertips, and not very forcefully. Still, there is no single behavior of deception and this certainly is not. You would be wise to merely consider this behavior and how it is performed

among other behaviors before you come to any conclusion about a person's honesty or sincerity.

332. **PULLING CLOTHING TO VENTILATE**—Pulling on the front of a shirt or other garment serves to ventilate the wearer. Whether the shirt is held out at the collar for a few seconds away from the neck or repeatedly plucked at and pulled away, this behavior serves to relieve stress, as do most ventilating behaviors. It is a good indicator that something is wrong. Obviously, in a hot environment, ventilating behaviors might simply be associated with the heat rather than stress. But remember, stress causes our temperature to increase, and this happens very quickly, which explains why in a difficult or testy meeting, people will be seen ventilating. Note that women often ventilate their dresses by pulling on the front top and midriff. Also of importance in a forensic setting is when a person ventilates as they hear a question or after they have answered it. Most likely they did not like the question.

333. **PLAYING WITH ZIPPER**—Playing with the zipper on a sweatshirt or jacket is a way of pacifying oneself when nervous or tense. Students might do this before a test if they are concerned and poker players do it also as they worry about their diminishing bank roll. Please note that it is a pacifying behavior and it can also be a way of dealing with boredom.

334. **LEANING AWAY**—Leaning away from a person is a form of distancing. If we are sitting next to someone who says something objectionable, we may subtly inch away from him. We often see this on talk shows. Rarely do we realize just how far away we lean from others when we find them disagreeable.

335. **SITTING BACK**—Pushing our chair back and leaning away from others at a table is in essence a distancing behavior that gives us additional insulation, so we can think and contemplate. Individuals who are unconvinced or still considering a point often will move slightly away until they are ready to engage, and then they will once again sit forward. For some it is a way to communicate they are taking themselves out for a few minutes to ponder this, or, and this is where other facial behaviors are useful, if they have decided they cannot support what is being discussed and so the pulling away is demonstrative of how they feel.

336. **SITTING FORWARD**—When we are ready to negotiate in good faith, or compromise, we tend to move from a leaning-back position to a sitting-forward position. This often telegraphs that we have made up our minds to move forward. One has to be careful, if sitting at a table or desk that is narrow, not to intimidate the negotiating partner by leaning too far forward. If negotiating with a team, make sure everyone is sitting

in the same way, and that eagerness to concede is not betrayed by someone on the team sitting forward before it is time to make it generally known.

337. **TURNING AWAY/VENTRAL DENIAL**—Our ventral or belly side is one of the most vulnerable places on the body. We will turn it away from others when we don't like them, they make us uneasy, or we don't like what they say. Upon meeting someone you don't care for, your facial greeting might be friendly but your belly will subconsciously shift away—what is called *ventral denial*—in essence denying that person your most vulnerable side. This can even take place among friends if something disagreeable is said. A good way to remember this: "Belly away don't want you to stay; belly away don't like what you say."

338. **BELLY/VENTRAL FRONTING**—When we like someone, we will turn our belly, or ventral side, toward her. We can see this behavior even in infants. It communicates that a person is interested and feels comfortable. When we meet someone while sitting down, if we like the person, we will, over time, reveal our shoulders and torso to that person as well. In summary, we show our gradual interest in others through ventral fronting.

339. **BELLY/VENTRAL COVERING**—The sudden covering of the belly with objects such as a purse or book bag

suggests insecurity or discomfort with what is being discussed. People will use everything from pillows (a couple arguing at home) to pets to their own knees to protect their ventral side when they feel threatened or vulnerable.

340. **POSTURAL ECHOING (MIRRORING)**—Our torso tends to echo the posture of those with whom we feel comfortable; this is called *isopraxis*. When standing with friends, people might find themselves mirroring one another's relaxed posture, a good sign that people are at ease together. In dating we will see one person lean forward, and the other, if comfortable, mimicking that behavior. Mirroring suggests agreement in conversation, mood, or temperament.

341. **RIGID SITTING**—A person who sits very rigidly without moving for long periods is undergoing stress. This is part of the freeze response, often seen in forensic settings, police interviews, and depositions, when people are so afraid, they can't move. The freeze response kicks in subconsciously, as if the person has just confronted a lion. Rigid sitting is not a sign of deception but rather an indicator of psychological discomfort.

342. **EJECTION-SEAT EFFECT**—A person in a stressful interview or who has been accused of something might sit in a chair as if ready to be ejected from a military

jet, gripping the armrests tightly. This, too, is part of the freeze response, and indicates deep distress or feeling threatened. What makes this behavior stand out is how rigid these individuals look, as if hanging on metaphorically for dear life.

343. **MOVING CHAIR AWAY**—This is a form of distancing when leaning away from others is just not enough. Literally, the person just moves the chair further and further back or away as if no one would notice. I have seen acrimonious discussions in academia where one professor moved completely away from the table to the corner of the room near the window—as if this were somehow normal. This behavior is motivated on a subconscious level to protect one's ventral side through distancing from a perceived threat, even if the threat is mere words or ideas.

344. **BODY SLOUCHING**—Slouching projects relaxation or indifference, depending on context. It is a perception-management technique often used by teenagers in dealing with their parents to demonstrate they don't care. In any formal professional setting, slouching should be avoided.

345. **DOUBLING OVER**—People in emotional turmoil might bend forward at the waist while seated or standing, as though experiencing intestinal distress. Usually they do

this with their arms tucked across their stomach. We see this behavior in hospitals and anyplace else where people might receive especially bad or shocking news.

346. **FETAL POSITION**—Under extreme psychological stress people might enter the fetal position. This is sometimes seen during intense arguments between couples, where one partner is so overwhelmed with emotions she will bring her knees up and sit in the fetal position—silent—to deal with the stress. She might also gather a pillow or some other object to hold against her belly (see #339).

347. **BODY CHILL**—Stress can cause individuals to feel cold in an otherwise comfortable environment. This is an autonomic response, in which the blood goes to the larger muscles, away from the skin, when we are threatened, stressed, or anxious, to prepare us to either run or fight.

348. **DRESSING THE TORSO**—Because our torso displays most of our clothing, it is important to mention here that clothing communicates and can give advantages to the wearer. Clothes often serve to project status within a culture. From name brands to the colors that we wear, clothing makes a difference in how we are perceived. It can make us more submissive or more authoritarian, or it can propel us into the job that we want. It can also

communicate where we are from or even where we are going as well as what issues we might be having. In every culture studied, clothes play a significant role. It is one more thing we must consider when we assess individuals to decode information they convey about themselves.

349. **BELLY COVERING DURING PREGNANCY**—Women often cover their suprasternal notch or throat with a hand when they feel concerned or insecure. But when they are pregnant, they will often raise their hand as if to go to the neck but then quickly move it to cover their belly, seemingly to protect their fetus.

350. **BELLY RUBBING**—Pregnant women will often repeatedly rub their belly to deal with discomfort, but also subconsciously to protect the fetus. Because it is a repetitive tactile behavior, it also serves as a pacifier and some researchers say it even helps to release oxytocin into the bloodstream.

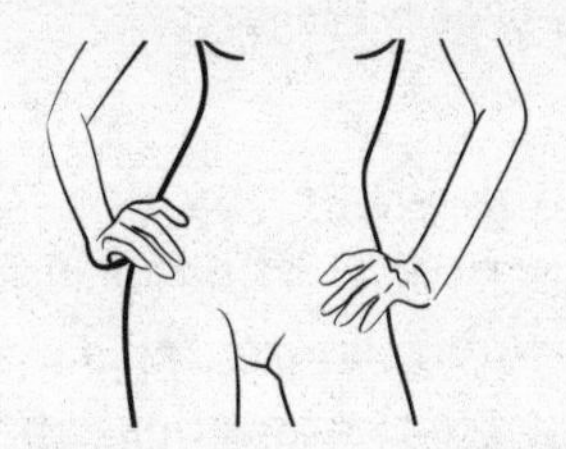

THE HIPS, BUTTOCKS, AND GENITALS

A body-language guide must include the area between the navel and the top of the legs. Our hips, angled just right so we can walk or run on two legs at great speed, give us shape and form, but they also say something about us—whether that something is about our reproductive health or our sensuality. As the renowned zoologist Desmond Morris points out in his book *Bodywatching,* throughout the world the hips and buttocks serve to attract and entice. The earliest sculpture ever discovered of a woman, the Venus of Hohle Fels, upward of 35,000 years old, is a masterpiece of the female form, emphasizing the hips, genitals, and the buttocks. Similar figurines have been found all over the world, which speaks to the natural attraction we have for this area of the body. Here is an opportunity to explore what this area of the body can communicate to us beyond the obvious.

351. **HIP/BUTTOCK SHIFTING**—Hip swiveling or buttock wiggling is a way of dealing with stress, boredom, or the fatigue of sitting in one place. People might also

do this during contentious debates, either when their emotions have been riled up or just afterward, as a process of calming down. You rarely see this with couples early on in their relationship; it tends to show up, if at all, down the road when there are issues being discussed.

352. **HIP RUBBING**—Under stress, people will rub the side of their hips and legs as a pacifier. It is also used to dry sweaty hands when feeling nervous. You see this gesture as students prepare to take a test or as travelers go through customs.

353. **HIP-TORSO ROCKING**—Individuals under psychological duress might rock forward and backward at the hips while sitting. Severe stress, such as witnessing the death of a loved one, will initiate this behavior, which pacifies with its repetitive motion. You might also see this behavior in people suffering from certain mental disorders such as those on the autism spectrum.

354. **HIP SWAYING**—When bored, we might find ourselves standing and swaying our hips side to side, as if cradling and rocking a baby to sleep. Swaying our hips causes the fluid and hairs in our inner ear to move, and that sensation is very soothing. This is different from hip-torso rocking (see #353), which is forward and back.

355. **HIP-OUT DISPLAY**—The hips are used by both men and women to get noticed, as in the famous statue of David by Michelangelo, where he stands contrapposto, with one leg slightly bent, making his buttocks more prominent and thus more attractive. Larger hips can also be used to garner attention—something Kim Kardashian does with pride and regularity. Hip-out displays are usually seen in courtship behavior to invite notice. In many cultures throughout the world the hips represent youth and fertility and are prominently displayed especially during the courtship years.

356. **GENITAL TOUCHING**—Teachers often report how young boys, and sometimes girls, will touch or pull at their genitals through their clothing. This is quite natural; the genitals contain an extraordinary number of nerve endings, which, when touched, not only soothe or calm children, but feels pleasant. Eventually kids outgrow this behavior, and it is not unusual or anything to be overly concerned about.

357. **CROTCH GRABBING**—This behavior, made famous by Michael Jackson while dancing, shocked many when first performed, yet today it is quite common in entertainers. There are many theories as to why some men do this: to garner attention, as a manhood display, or merely to adjust for comfort. On adult males, it can be quite disturbing if done repetitively and at a close

distance such as an office, as women have reported to me. It certainly should be avoided in public.

358. GENITAL FRAMING—Often done by cowboys in movies or in photographs, this is where the man places the thumbs inside the pants or hooks them on his belt and the fingers straddle the crotch area. Genital framing is used to garner attention and serves as a masculine display. Usually the elbows are thrust outward, making the man look bigger and appear tougher.

359. GENITAL COVERING—We might place our hands together over our genitals or crotch in certain situations—in elevators, for example, where men will look at the numbers or the door while doing so. This behavior can be effective in dealing with social anxiety or if someone is standing too close.

360. SITTING KNEES FAR APART—Often referred to as "manspreading," this is where a man sits, often on public transportation, with the knees far apart. This is considered rude due to how much space a person takes up in this position, and the less-than-diplomatic display of the inner legs and the crotch, and it is inconsiderate.

THE LEGS

Our legs are unique in the animal kingdom as they point inward at the hips, allowing us to walk, run, sprint, climb, kick, hurtle, swim, and bicycle. We use our legs for locomotion, for protection, to establish dominance, and as a steady anchor for our children to hang on to when they are nervous or shy. Sinewy, long, or stocky, legs are as varied as their owners. Often ignored when it comes to nonverbals, they can communicate everything from elegance to nervousness to joy. And because our legs serve as a survival tool—they can help us escape—they can be very honest when it comes to how we feel about others.

361. SPATIAL DISTANCING—The anthropologist Edward T. Hall coined the term *proxemics* to describe the need all animals have for personal space. If someone stands too close to us, we are uncomfortable. Our spatial needs are based on both culture and personal preference. Most Americans feel comfortable in public spaces at a distance of 12 to 25 feet from others; in social spaces 4 to

12 feet is preferred; while our personal space is comfortable at about 1.5 to 4 feet. When it comes to our intimate space, anything less than a foot, we are very sensitive to who gets that close. These are of course approximations, as it is different for everyone and varies with culture, nationality, location, and even time of day. At night we might not feel comfortable walking near a stranger who is closer than ten feet.

362. TERRITORIAL STANCE—We use our legs as a form of territorial display by how we stand. The further apart our feet, the greater the territorial display. The breadth of a person's stance is telling: military men and police officers tend to stand with their feet farther apart than, say, accountants and engineers. The spread of the legs transmits clearly a sense of confidence and a subconscious claiming of territory.

363. TERRITORIAL CHALLENGE—During a heated argument a person might intentionally invade your personal space, getting just inches away from your face (figuratively "in your face"), puffing out the chest, and glaring. This violation of space serves to intimidate, and might be a prelude to a physical assault.

364. ANGLING TO THE SIDE—Most people prefer to talk to others from a slightly angled position, rather than directly face-to-face. When children first meet, they usually approach each other at angles for a reason—they get a better reception. I have found that when businesspeople stand facing each other at a slight angle, the amount of time they spend together increases. Note that when there is acrimony, it is always best to stand angled slightly away from the other person as this tends to help diffuse negative emotions.

365. WALKING BEHAVIORS—The way we walk communicates a lot. Some walks are intentionally sexy (Marilyn Monroe's for example), while others show strength and determination (John Wayne's). Some walks suggest that a person is on an important task, while others are more relaxed and casual, or intended to get a person noticed, like John Travolta's character's walk in the opening sequence of the movie *Saturday Night Fever*. And it is not just how we walk, sometimes we communicate our interest in others by how frequently we walk by to get a good look or to get noticed.

366. SETTING PACE DURING WALK—Whoever sets the walking pace in a group is usually the person in charge. We will speed up or slow down for the most senior person or group leader. Even teenagers will do this, deferring to the most socially prominent one among them by walking at that person's pace. This might mean the last person in a group is the leader and is setting the pace to walk no faster. In analyzing groups remember that it is not who is in front but rather who sets the pace that is in charge.

367. SITTING BEHAVIORS—Each culture sits differently. In some parts of Asia, people squat, buttocks low and knees high, while waiting for a bus. In other cultures, the legs are intertwined as you sit, as Gandhi did while working a loom. In Europe and elsewhere, people often sit with one leg draped over the opposite knee so that the sole points downward. In America, you will see a combination of sitting styles, including the figure 4, where the ankle is placed on top of the opposite knee, with the foot noticeably high. When it comes to sitting behaviors, it is important to follow both local customs as well as those of your host.

368. HOLDING LEGS TOGETHER, SITTING—Our level of confidence is often revealed by how we sit. Legs that suddenly come together suggest insecurity. In part, of course, the way we sit is cultural, but some people will move their legs with great reliability depending on how they feel emotionally, revealing their degree of self-assurance. Keep in mind that in many places, women will sit with their knees together as a matter of social convention.

369. LEGS SPREADING APART, SITTING—Legs that are suddenly set wider apart while sitting during an interview or a conversation suggest greater comfort or confidence. This is a universal territorial display; the farther apart they are, the more territory is being claimed. This behavior is more pronounced in men.

370. **ANKLES LOCKING**—While sitting down, especially in a formal setting, people will often bring their ankles together and interlock them. I look for people who suddenly perform this gesture when something controversial or difficult is being discussed; it usually signals that they are restraining themselves, expressing reservation, or showing hesitation or psychological discomfort.

371. **ANKLES INTERLOCKING AROUND CHAIR LEGS**—Insecurity, fear, or concern will cause some people to suddenly interlock their ankles around the legs of a chair. Some people, of course, sit like this routinely. However, the suddenly interlocking of ankles around the chair following a question, or while discussing a sensitive issue, is a strong indicator that something is wrong. It's part of the freeze/self-restraint response.

372. **KNEE CLASPING, LEANING BACK**—A firm knee clasp can signify that a person is self-restraining. You often see this among nervous job applicants. The feet are on the ground, the knees tightly clasped, and because of stress, the person is leaning back rather rigidly.

373. KNEE CLASPING, LEANING FORWARD—When we do this from the sitting position, hands on knees, leaning forward, it usually means we are ready to leave. Often we will also align the feet in the starter's position, one in front of the other. Do not do this in a meeting unless you are the senior-most person; it is insulting to signal that you want to leave if someone else is in charge or superior to you.

374. CROSSING LEG AS BARRIER, SITTING—Crossing a leg in such a way that it acts as a barrier while sitting—with the knee high over the opposite leg—suggests that there are issues, reservations, or social discomfort. Whether at home or at work, this behavior accurately reflects feelings. You often see this occur the instant an uncomfortable topic is brought up.

375. LEG DRAPING—Subconsciously, individuals who feel confident or superior will drape their leg over a desk, chair, or object—even other people—as a way of establishing a territorial claim. Some bosses do this regularly, sitting in one chair and draping their leg over another.

376. LEG RUBBING—Rubbing the tops of our quadriceps—a gesture known as a *leg cleanser*—works to pacify us when we are under high stress. It can be easy to miss, since it usually occurs under a table or desk.

377. **KNEE RUBBING**—We see repeated scratching or rubbing of the area just above the knee in people who are feeling stress or anticipating something exciting. Like most repetitive behaviors, it serves to pacify, assuaging the excitement or tension.

378. **ANKLE SCRATCHING**—In tense situations it is not unusual for a person to scratch at the ankles. It serves both to relieve stress and to ventilate the skin. We often see this in high-stakes situations such as a large pot in a poker game or in a forensic interview when a difficult question is asked.

379. **KNEE FLEXING**—This behavior is performed by quickly flexing the knees forward while standing, which causes the person to sink down rather noticeably. Usually the person immediately recovers. This is a very juvenile behavior, almost akin to the beginnings of a temper tantrum. I have seen grown men do this at the car-rental counter when told the car they requested is not available.

380. **DRAGGING FEET**—We often seen children drag their feet back and forth while talking or waiting for something. This is a repetitive behavior that helps them calm down or pass idle time. Adults might do it as they await someone's arrival. It can be used to mask anxiety and is a common behavior with shy inexperienced people on a first date.

381. ANKLE QUIVERING—Some people while standing will repetitively twist or quiver their foot to the side at the ankle, in a show of restlessness, agitation, animosity, or irritation. This is very perceptible because the shaking causes the whole body to move.

382. KNEE-HIGH SELF-HUGGING—We often see teenagers hug their own legs by bringing the knees up to chest level. This can be very comforting and helps them enjoy a moment as they listen to music or to help them deal with emotions. I have also seen some criminals do this while being interviewed to help deal with stress.

383. STANDING LEG CROSSING (COMFORT)—We cross our legs while standing when we are alone or if we feel comfortable with the people around us. The minute someone causes us the slightest psychological discomfort, we will uncross the legs in case we need to quickly distance or defend ourselves from the offending person. You may notice this in elevators, where a lone rider will uncross the legs the minute a stranger enters.

384. LEG KICKING, SITTING—A leg crossed over the knee that goes from shaking or twitching (repetitive movement) to sudden kicking up and down after a question is asked indicates high discomfort with the question.

This is not a pacifier, unless the person does it all the time. It is a subconscious act of kicking away something objectionable. Sudden leg kicks, in response to a question or a statement, are usually associated with strong negative feelings.

385. **JUMPING (JOY)**—Positive emotions drive this gravity-defying behavior that is displayed around the world. Primates also will jump for joy, much like humans do, when they sense they are about to get a treat. Our limbic system, the emotional center of the brain, directs this behavior automatically, which is why when a player scores a point spectators jump up all at once, without being told to.

386. **UNCOOPERATIVE LEGS AND FEET**—Children and sometimes adults will protest with their feet by dragging them, kicking, twisting, or going dormant turning themselves into a dead weight. Children do this when they refuse to go someplace they don't want to go to and often you will see adults peacefully resisting arrest doing the same. Their legs are clearly and unequivocally demonstrating how they truly feel about something.

387. **LOSING ONE'S BALANCE**—There are any number of medical conditions that can trigger loss of balance, including low blood pressure, or something so simple as

getting up too quickly. Drugs and alcohol might also play a role. Age can be a factor as well. When we see someone suddenly lose his balance, our first instinct must be to assist where possible. It is important to note that when the elderly fall, it can have catastrophic consequences due to frail bones and so immediate action is required.

THE FEET

"The human foot is a masterpiece of engineering and a work of art," said Leonardo da Vinci after decades of dissecting and studying the human body. Though relatively small compared with other parts of the body, the feet carry our full weight and are invaluable in sensing motion, vibrations, heat, cold, and humidity. We put more pressure on our feet than on any other part of our bodies, and we punish them with tight shoes and endless journeys. Sensitive to the slightest touch, they can be very sensual—or they can break a brick with a karate kick. Like the rest of the body, they do their intended job exquisitely, balancing us, allowing us to walk and climb, but they also communicate our feelings and intentions as well as our fears.

388. FROZEN FEET—Feet that suddenly go "flat" and stop moving indicate concerns or insecurities. We tend to freeze movement when we are threatened or worried, an evolutionary response that keeps us from being noticed by predators.

389. FOOT WITHDRAWING—During job interviews, interviewees will suddenly withdraw their feet and tuck them in under their chairs when they are asked sensitive questions they might not like. The movement is sometimes rather noticeable, closely following a question that is difficult to answer, such as "Have you ever been fired from a job?" At home, teens might do this when asked where they were the night before.

390. PLAYING FOOTSIES—When we like another person, our feet will be drawn to them. When we like them romantically, our feet might move almost subconsciously toward theirs so that they come into contact. This is why you see people playing footsie under the table in the early stages of a relationship. The playful touching has an important role in connecting us to others. Neurologically, when our feet are touched, it registers on a sensory strip along the parietal lobe of the brain, very close to where our genitals also register.

391. FOOT ROCKING—This is another repetitive behavior that serves to pacify us. We might do this when we're waiting for someone to hurry up—the rocking shifts from the heels to the toes, back and forth. Since this elevates us as we rock forward, it is also somewhat of a gravity-defying behavior. Foot rocking can both alleviate boredom and demonstrate that a person is in charge.

392. **FOOT TURNING AWAY**—When we're talking to someone, we might signal that we need to leave by gradually or suddenly pointing one foot toward the door. This is our nonverbal way of communicating "I have to go." It is an *intention cue,* and if the person we're talking to ignores it, we can become very irritated. Be mindful of others, and recognize that when their foot turns away, chances are they have to go.

393. **FEET TURNING AWAY**—When we are in the presence of someone we dislike, it is not unusual for our feet to turn together toward the door or away from that person. In watching juries over the years, I have noted that jurors often turn their feet toward the jury room the instant a witness or attorney they dislike begins to speak. At parties, you might see two people look at each other and even exchange a social smile while their feet will turn away, indicating their mutual dislike.

394. **TOES POINTING INWARD/PIGEON TOES**—Some people turn their toes inward (sometimes called "pigeon toes") when they are insecure, shy, or introverted, or when they feel particularly vulnerable. This

behavior, which is generally seen in children but also in some adults, demonstrates some sort of emotional need or apprehension.

395. **TOES POINTING UP**—Occasionally, when someone is engaged in conversation, either in person or over the phone, you will see the toes of one foot point up, at an angle, with the heel firm to the ground. This is a *gravity-defying behavior,* which is usually associated with positive emotions. When good friends run into each other, you will also see this behavior as they talk.

396. **EXPOSING SOLES OF FEET**—In many parts of the world, especially the Middle East, Africa, and parts of Asia, displaying the sole of one's foot or shoe is insulting. When traveling abroad, be careful how you sit—resting the ankle on a knee exposes your soles. It is usually preferable to either keep both feet on the ground or to drape one leg over the opposite knee so that the sole is pointing downward.

397. **BOUNCY HAPPY FEET**—We sometimes register an emotional high with happy feet—the feet are animated and jumpy. This is certainly visible in children, when you tell them you're taking them to a theme park, for example. But we also see it in adults. Poker players, for instance, might bounce their feet under

the table when they have a monster hand. While the feet themselves might not be visible, often they will cause the clothing to shake or tremble all the way up to the shoulders.

398. **FOOT TAPPING**—This is a familiar behavior used to pass the time, to keep tempo with music, or, like finger strumming, to indicate that we are becoming impatient. Usually just the front of the foot is involved, while the heel remains grounded, but it can also be done with the heel of the foot.

399. **TOE WIGGLING**—Ever find yourself wiggling your toes? Chances are you were feeling good about something, excited, or eagerly anticipating an event. The movement of the toes stimulates nerves that help to alleviate boredom or stress and can signal excitement in much the way happy feet do.

400. **AGITATED FEET**—Every parent recognizes the agitated feet of a child who wants to leave the table to go play. Often our feet will telegraph that we want to leave, even in a boardroom full of adults, through excessive uncomfortable movements. These might include repetitive shifting, movement from side to side, foot withdrawal, or repetitively raising and lowering the heels of the feet.

401. NERVOUS PACING—Many people will pace when they are stressed. This acts as a pacifier, as all repetitive behaviors do.

402. LEGS AS INDICATORS OF DESIRES—Our legs often signal when we want to get closer to something or someone. Legs and feet will gravitate toward a store window displaying candy, or a person we are interested in. Or we might lean away as if to leave but our legs remain frozen in place because we like the person we are with.

403. LEG TANTRUMS—These are most often seen in children when they twist, move, and energetically stomp their legs, letting everyone know how they feel. And it's not just children, from time to time you will see adults do the same, as I did when an executive was bumped from a flight. This is a reminder that the legs also demonstrate emotions, and because these are the largest muscles in the body, they do so with maximum effect.

404. FOOT STOMPING—Children are not the only ones who stomp their feet to make their feelings known. We often see this when people are exasperated or they have reached the limit of their patience. I've observed men and women stomping their feet in lines that move too slowly. Usually the foot is only stomped once, just to get noticed.

405. SOCK PULLING—Stress will cause skin temperature to rise quickly. For many people, their feet and lower legs become uncomfortably warm. When stressed, they will ventilate their ankles by pulling on their socks, sometimes repeatedly. This is an often unnoticed behavior that signals a high degree of psychological discomfort.

406. SHOE DANGLE—When some people, especially women, are comfortable around others, they will dangle their shoe near the instep of the foot. This is often seen in dating situations. The shoe will be slipped back on the very instant a woman feels uncomfortable or no longer likes what the other person is saying.

407. GENERAL FOOT AND LEG AGITATION—A person may present in an agitated state wherein their feet are restless and they shift or pace, racing to and fro seemingly without purpose. This might be because of a diagnosable event, such as an allergic reaction to a drug, illicit drug use, shock after a tragedy, or a panic attack. Concurrently, they might display clenched fists, fidgety hands, some lip biting, and even eye twitching. This generalized state of agitation is a nonverbal signal that something is wrong and the person is struggling to deal with it. Medical assistance or psychological counseling might be needed. Don't expect the person experiencing such agitation to be able to speak or think coherently at a moment like this.

CONCLUSION

My hope for this book is that it will open your eyes to the world around you, to help you understand and appreciate others through this unspoken language we call nonverbals. But reading about it is only the first step. Now comes the more interesting part: looking for and testing what you have learned. By verifying these observations on your own, "in the field," every day, you will develop your own skill set for decoding human behavior. The more you study and verify, the easier it becomes, and you will come to immediately notice signs others miss.

We humans are all in the people business. To be attuned to others is to care. Leadership is all about understanding and communicating, and body language is a key piece of that. Effective leaders listen and transmit on two channels: the verbal and the nonverbal. And even though our world is becoming increasingly digitized and depersonalized, face-to-face contact is still extraordinarily important in building relationships, establishing trust and rapport, understanding others, and relating empathetically. Technology has its uses—it helped me write this book—but it has limitations when it comes to selecting a best friend or someone to spend your life with. The nonverbal cues we give, and those we notice in others, matter significantly.

Of course, no book can encompass all of human behavior. Others will focus on different behaviors and contribute to our knowledge beyond my scope—perhaps one day it will be you. It has been my intent always to share my knowledge and experiences with others, and doing so has brought me great happiness. I hope you also will share with others what you've learned about body language and nonverbal communication. May your life be as enriched as mine has been, learning about why we do the things we do. It has been an interesting ride. Thank you for sharing it.

ACKNOWLEDGMENTS

I begin each journey into writing fully aware and mindful that so many people have helped me along the way and not just in writing. Most will never be recognized because I have long forgotten the name of a teacher who answered a question, or the neighbor who shared a lunch, or the coach who taught me to discipline my focus. I have forgotten their names but not their acts of kindness. Nor have I forgotten the countless people all over the world, from Beijing to Bucharest, who have honored me by buying my books, following me on social media, and encouraging me to write. A hearty thank-you.

To Ashleigh Rose Dingwall, thank you for your assistance in reading the manuscript and for your valuable suggestions. To the men and women of the FBI, especially those in the prepublication review unit, thank you for your tireless assistance always.

William Morrow is presently home to four of my books precisely because of people like publisher Liate Stehlik and the wonderful team who worked on this project including Ryan Curry, Bianca Flores, Lex Maudlin, and production editor Julia Meltzer. To my editor at William Morrow, Nick Amphlett, who championed this project, expertly guiding it through its many paces, I have more than gratitude. Nick, you were most kind and generous with your time, your ideas, and the editing process. You and your colleagues collectively made this work possible and I thank you.

To my dear friend and literary agent Steve Ross, director of the Book Division at the Abrams Artist Agency, you have my most profound gratitude. Steve is the kind of agent most writers wish they had because he listens, he cares, he counsels, and he gets things done. Steve, you are unique. Thank you for your guidance and leadership when it was needed the most. A big thank-you also goes out to your colleagues David Doerrer and Madison Dettlinger for their assistance on this and other projects.

I would not be here writing if not for my family, who have always supported me and allowed me to be curious and follow my own path less taken. To Mariana and Albert, my parents, thank you for all the sacrifices you made so that I could triumph. To my sisters, Marianela and Terry, your brother loves you. To Stephanie, my daughter, you have the loveliest of souls. To Janice Hillary and my family in London, thank you for your encouragement and understanding—always.

Lastly to my wife, Thryth, who is so wonderfully supportive of everything I do, but especially of my writing—

thank you. From your kindness I draw strength and from your encouragement I aspire to be better in all things. I am a far better person since you entered my life. Your love is felt each day in the most important of ways—by everything you do.

BIBLIOGRAPHY

Alford, R. (1996). "Adornment." In D. Levinson and M. Ember (Eds.), *Encyclopedia of Cultural Anthropology*. New York: Henry Holt.

Burgoon, J. K., Buller, D. B., & Woodall, W. G. (1994). *Nonverbal communication: The unspoken dialogue*. Columbus, OH: Greyden Press.

Calero, H. H. (2005). *The power of nonverbal communication: How you act is more important than what you say*. Los Angeles: Silver Lake Publishers.

Carlson, N. R. (1986). *Physiology of behavior* (3rd ed). Boston: Allyn & Bacon.

Darwin, C. (1872). *The expression of emotion in man and animals*. New York: Appleton-Century Crofts.

Dimitrius, J., & Mazzarela, M. (1998). *Reading people: How to understand people and predict their behavior—anytime, anyplace*. New York: Ballantine Books.

Ekman, P., Friesen, W. Y., & Ellsworth, P. (1982). *Emotion in the human face: Guidelines for research and an integration of findings*. Ed. Paul Ekman. Cambridge, UK: Cambridge University Press.

Etcoff, N. (1999). *Survival of the prettiest: The science of beauty*. New York: Anchor Books.

Givens, D. G. (2005). *Love signals: A practical guide to the body language of courtship*. New York: St. Martin's Press.

———. (1998–2007). *The nonverbal dictionary of gestures, signs & body language cues*. Spokane, WA: Center for Nonverbal Studies. Http://members.aol.com/nonverbal2/diction1.htm.

———. (2010). *Your body at work: A guide to sight-reading the body language of business, bosses, and boardrooms*. New York: St. Martin's Press.

Hall, E. T. (1969). *The hidden dimension*. Garden City, NY: Anchor Books.

———. (1959). *The silent language*. New York: Doubleday.

Iacoboni, M. (2009). *Mirroring people: The science of empathy and how we connect with others*. New York: Picador.

Knapp, M. L., & Hall, J. A. (2002). *Nonverbal communication in human interaction* (5th ed.). New York: Harcourt Brace Jovanovich.

LaFrance, M., & Mayo, C. (1978). *Moving bodies: Nonverbal communications in social relationships*. Monterey, CA: Brooks/Cole.

LeDoux, J. E. (1996). *The emotional brain: The mysterious underpinnings of emotional life*. New York: Touchstone.

Montagu, A. (1986). *Touching: The human significance of the skin*. New York: Harper & Row Publishers.

Morris, D. (1985). *Bodywatching: A field guide to the human species*. New York: Crown Publishers.

———. (1994). *Bodytalk: The meaning of human gestures.* New York: Crown Trade Paperbacks.

———. (1971). *Intimate behavior.* New York: Random House.

———. (1980). *Manwatching: A field guide to human behavior.* New York: Crown Publishers.

———. (2002). *Peoplewatching: A guide to body language.* London: Vintage Books.

Morris, Desmond, et al. (1994). *Gestures.* New York: Scarborough Books.

Navarro, J. (2016). "Chirality: A look at emotional asymmetry of the face." *Spycatcher* (blog). *Psychology Today,* May 16, 2016. https://www.psychologytoday.com/blog/spycatcher/201605/chirality-look-emotional-asymmetry-the-face.

Navarro, J., & Karlins, M. (2007). *What Every BODY Is Saying: An ex-FBI agent's guide to speed-reading people.* New York: HarperCollins Publishers.

Navarro, J., & Poynter, T. S. (2009). *Louder than words: Take your career from average to exceptional with the hidden power of nonverbal intelligence.* New York: HarperCollins Publishers.

Panksepp, J. (1998). *Affective neuroscience: The foundations of human and animal emotions.* New York: Oxford University Press.

Ratey, J. J. (2001). *A user's guide to the brain: Perception, attention, and the four theaters of the brain.* New York: Pantheon Books.

INDEX

ALSO BY JOE NAVARRO

THE DICTIONARY OF BODY LANGUAGE
A Field Guide to Human Behavior

"*The Dictionary of Body Language* stands out as a clear, simple, and accurate field guide to nonverbal communication and human behavior. Keep this book handy when investigating any human interaction for yourself, and you'll be amazed by the clarity and insight it will bring to you."

—Mark Bowden, bestselling co-author of *Truth and Lies: What People Are Really Thinking*

LOUDER THAN WORDS
Take Your Career from Average to Exceptional with the Hidden Power of Nonverbal Intelligence

"*Louder Than Words* takes us from an understanding of nonverbal behavior to an understanding of something far more valuable for success—nonverbal intelligence."

—Robert B. Cialdini, author of *Influence: Science and Practice*

WHAT EVERY BODY IS SAYING
An Ex-FBI Agent's Guide to Speed-Reading People

"A masterful work on nonverbal body language by an exceptional observer. Joe Navarro's work has been field-tested in the crucible of law enforcement at the highest levels within the FBI. I cannot praise the book enough."

—David Givens, Ph.D., author of *Crime Signals* and *Love Signals*